AF470907

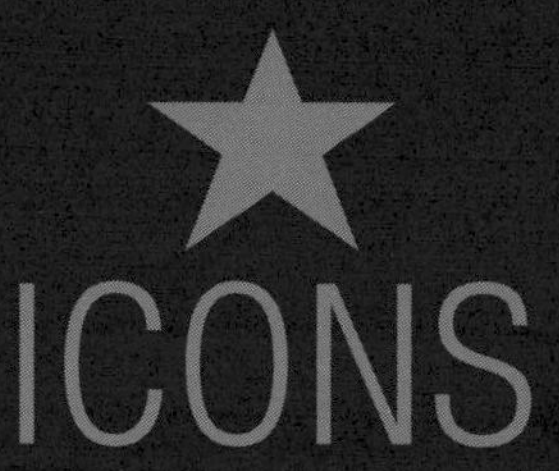

Robots and Spaceships

W27
£2
M32

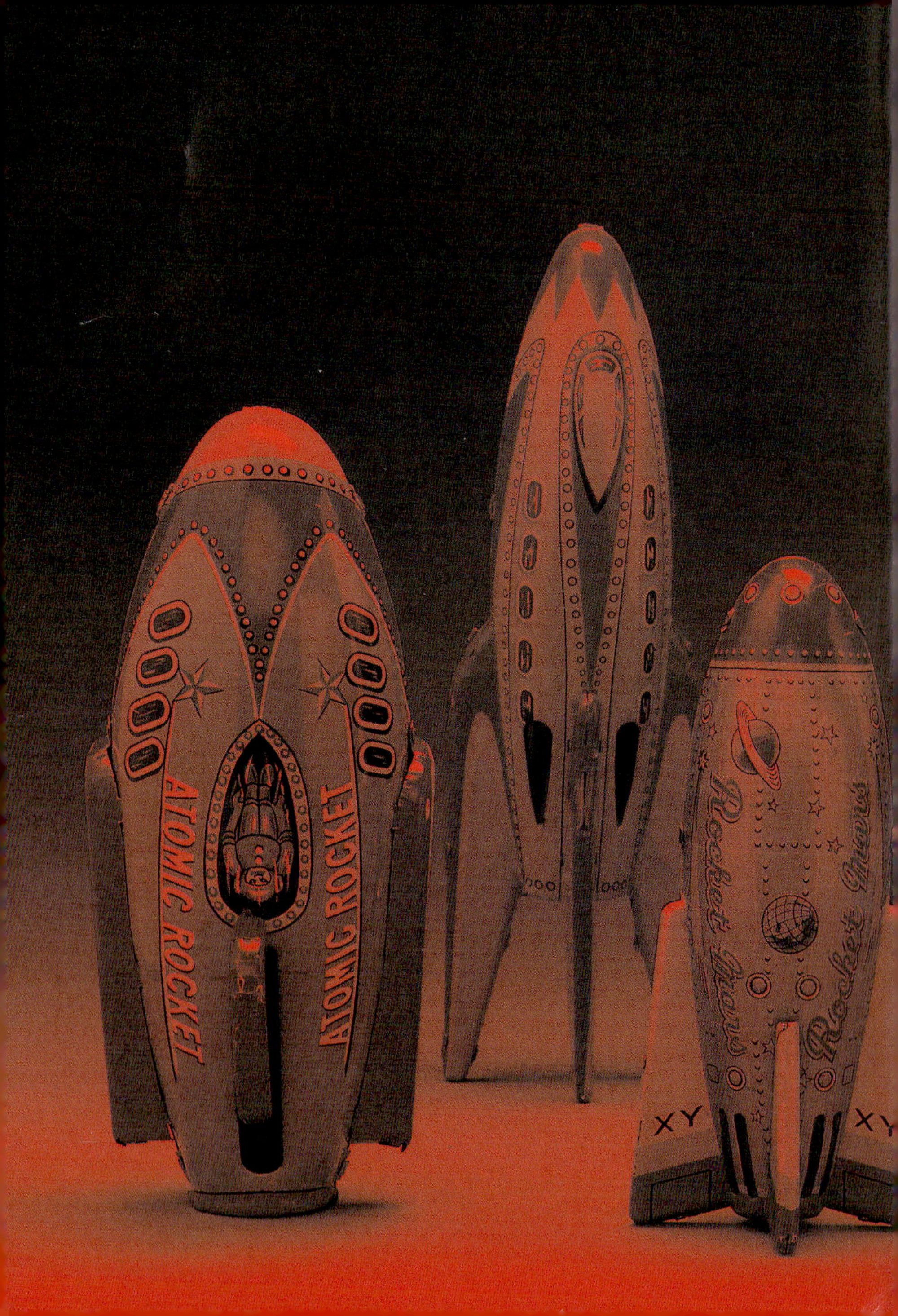

ATOMIC ROCKET
ATOMIC ROCKET
Rocket Mars
Rocket Mars
XY
XY

Teruhisa Kitahara
Yukio Shimizu

Robots and Spaceships

TASCHEN

KÖLN LONDON LOS ANGELES MADRID PARIS TOKYO

Front and back cover:
1960s, *Moon Scout*, Louis Marx, 90 x 160 x 290 mm

The captions contain the following information:
Date of production, title, manufacturer, size
(depth, width, height)

Die Bildlegenden enthalten folgende Angaben:
Entstehungszeit, Titel, Hersteller, Maße
(Tiefe, Breite, Höhe)

Les légendes donnent les indications suivantes:
Date de fabrication, titre, marque, dimensions
(profondeur, largeur, hauteur)

To stay informed about upcoming TASCHEN titles,
please request our magazine at www.taschen.com or write
to TASCHEN, Hohenzollernring 53, D–50672 Cologne,
Germany, Fax: +49-221-254919. We will be happy to send
you a free copy of our magazine which is filled with
information about all of our books.

© 2001 TASCHEN GmbH
Hohenzollernring 53, D–50672 Köln
www.taschen.com

© for the illustrations: Teruhisa Kitahara;
Photos by Yukio Shimizu
Design: Burkhard Riemschneider, Cologne
Cover design: Angelika Taschen, Cologne
Editorial coordination: Shibata Asuka, Tokyo;
Ute Kieseyer, Cologne
English translation: Yuko Aoki, Tokyo
French translation: Daniel Roche, Paris
German translation: Christiane Schmieger, Cologne

Printed in Italy
ISBN 3–8228–5566–9

Contents

Inhalt

Sommaire

A Passion for Robots

As a child in the mid-1950s and early 1960s, I would make believe I was a robot or an astronaut, in a world of booming, clanking machines and rockets. I grew up surrounded by movies, comics, and stories brimming with science-fiction themes. In those days of early space travel and the Apollo program, toyshops were full of robots and rockets. Science was the stuff that dreams were made of.

Whenever I got a new toy based on one of the many movie heroes or TV characters of the day, I would play with it all day long. If a toy got broken, I would open it up and inspect the inside with the precision of a surgeon, scrutinising the gears, coils, meters, and springs. As I grew older, my favorite robots became worn and torn, and were eventually thrown away.

Fifteen years on, as an adult, I still recall the shock of discovering toy robots again. It was a strange feeling of excitement, something stronger than mere nostalgia for the playthings that had once fired my childish imagination.

One of the curious attractions of rediscovering toy robots as an adult lies in noticing the color and texture of the material, and the eye for detail with which they were made. Most of them, produced in the 1950s and 1960s, are labeled Made in Japan. In fact, in those days, most Japanese products were exported to the States in packages with English labeling that still seem modern. I never fail to be astonished at the idea of toy companies investing so much passion and technology in creating a toy space station computer control centre, let alone a radio-controlled robot. It is probably this design approach that makes them more than mere toys, giving them a solidity and material presence that comes across even in two dimensional graphics and photographs. And this probably explains why my robots are so often used for advertisements and record covers.

Another interesting aspect of toy robots is the number of variations in each period. For example, when television was introduced, TV-shaped toys were developed. When Apollo landed on the Moon, space robots, astronauts and space stations were produced. Also, popular characters such as Robby the robot from the 1956 science-fiction blockbuster "Forbidden Planet" was manufactured in more than 300 different versions.

When I was just starting my collection, fascinated by the sheer vitality of the genre, I came across a French book on robots at a foreign bookstore in Tokyo. I was astonished to find that there were even people in faraway countries who shared my interest. Today, of course, there are many more collectors in the USA, where toy robots change hands at higher prices. I have the feeling that robots have become a source of international communication, with a world-wide appeal.

Robots have always been a human dream, yet futuristic as they are, these toys bear all the signs of skilled mid-century craftsmanship. I cannot tell whether man is in thrall to his machines, but I do know that many adults are still entranced by toy robots.

Many technological advances that were once confined to the realm of dreams are now within our grasp and more are yet to come as the century progresses. Little did we imagine, back in those days when a pencil stuck into a pencil case became a space-age walkie-talkie, that children would soon be seen chatting on their mobile phones at every street corner. In much the same vein, it was accepted that robots would never be able to move in a smooth or fluid way. Today, people even have robot pets instead of real ones, and "dancing" robots are manufactured for the home. However low-tech these toy robots may appear to contemporary eyes, it is precisely this that has made them so popular in a high-tech world that is rapidly losing its lustre.

Toys discarded in the 20th-century era of mass production and mass consumption have survived and been added to my collection, and are now enjoying a second lease of life. I hope that younger generations, who have never had the pleasure of actually holding one of these toy robots in their hands, will nevertheless sense the warmth they exude in my seven museums or in this book. I believe that the mysterious charm of these robots will radiate all the more in the spirit of the 21st century.

Teruhisa Kitahara

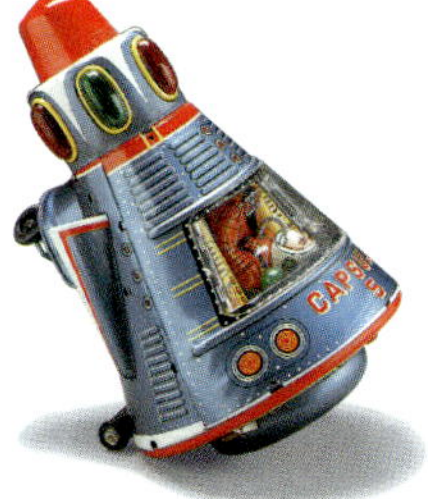

Roboter – meine Leidenschaft

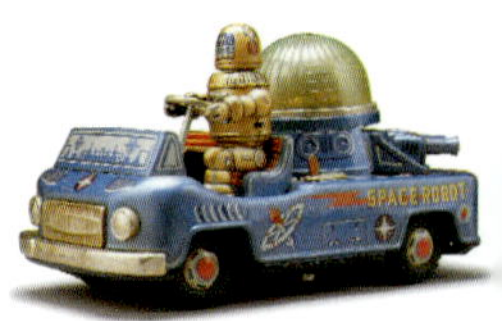

Mitte der 1950er und zu Beginn der 1960er Jahre, als auf der ganzen Welt lärmende
Maschinen und Raketen wie Pilze aus dem Boden schossen, spielte ich als Kind häufig
Roboter oder Astronaut. Ich wuchs auf inmitten von Filmen, Comics und Geschichten,
die alle von Sciencefiction handelten. In jenen Tagen, als die ersten Menschen in den Welt-
raum reisten und das Apollo-Programm an den Start ging, steckten die Spielzeugläden
voller Roboter und Raketen. Die Menschen träumten von der Wissenschaft.

Immer wenn ich ein neues Spielzeug erworben hatte, das einem der zahlreichen Film-
oder Fernsehhelden jener Zeit nachempfunden war, spielte ich den ganzen Tag lang damit.
Ging ein Spielzeug kaputt, untersuchte ich sein Inneres mit der Präzision eines Chirurgen,
wobei ich sämtliche Zahnräder, Spiralen, Zähler und Federn genauestens überprüfte. Je
älter ich wurde, desto mehr nutzten sich meine liebsten Roboter ab, so dass sie schließlich
weggeworfen wurden.

Fünfzehn Jahre später entdeckte ich die Roboter wieder. Noch heute erinnere ich mich
sehr gut daran: Ich war seltsam aufgeregt, aber dieses Gefühl war mehr als bloße Nostalgie
angesichts all der Sachen, die früher meine wildesten Fantasien angeregt hatten.

Wenn man die Spielzeugroboter als Erwachsener wiederentdeckt, üben in erster Linie
die Farben und Strukturen des Materials eine merkwürdige Anziehungskraft aus und man
achtet ganz besonders auf die Details ihres Aufbaus. Sie entstanden vor allem in den 1950er
und 1960er Jahren und tragen die Aufschrift „Made in Japan". Damals wurden die meisten
japanischen Produkte in die Vereinigten Staaten exportiert. Die Verpackungen besaßen eng-
lische Beschriftungen, die auch heute noch zeitgemäß wirken. Ich staune immer wieder,
wie viel Leidenschaft und technischen Aufwand die Spielzeughersteller zum Beispiel in ein
Computer-Kontrollzentrum einer Spielzeug-Raumstation oder einen funkgesteuerten Roboter gesteckt haben. Wahrscheinlich ist es gerade diese Auffassung vom Design, von der
Konstruktion, die die Dinge über das bloße Spielzeug-Dasein erhebt und ihnen eine materi-
elle Präsenz verleiht, die sogar in zweidimensionalen Abbildungen und Fotos noch spürbar
ist. Das könnte auch erklären, warum meine Roboter bevorzugt für Werbezwecke und die
Gestaltung von Plattencovern verwendet werden.

Ein weiterer interessanter Aspekt von Spielzeugrobotern ist die Vielfalt ihrer Erschei-
nungsformen zu bestimmten Zeiten. In der Anfangszeit des Fernsehens zum Beispiel wurden fernsehförmige Spielzeuge produziert. Als die Apollo auf dem Mond
landete, entstanden Weltraumroboter, Astronauten und Raumstationen.
Auch beliebte Figuren wie Robby, der Roboter aus dem Sciencefiction-

Film „Alarm im Weltall" (Forbidden Planet) von 1956, kamen in mehr als 300 verschiedenen Varianten auf die Welt.

Als ich gerade erst mit dem Sammeln begonnen hatte und von der Lebendigkeit des Genres einfach hingerissen war, entdeckte ich in einem ausländischen Buchladen in Tokio ein französisches Buch über Roboter. Ich war erstaunt, dass es selbst in fernen Ländern Leute gab, die dasselbe Interesse verfolgten wie ich. Heutzutage gibt es in den Vereinigten Staaten natürlich noch viel mehr Sammler und die Spielzeugroboter wechseln für sehr viel mehr Geld ihre Besitzer. Mittlerweile habe ich den Eindruck, dass die Roboter zu einer attraktiven Quelle internationaler Kommunikation geworden sind.

Roboter gehörten schon immer zu den Träumen der Menschheit. Und obwohl diese Spielzeuge eher futuristisch wirken, zeugen sie von der hochwertigen Handwerkskunst aus der Mitte des vergangenen Jahrhunderts. Schwer zu sagen, ob der Mensch zum Sklaven seiner Maschinen geworden ist – fest steht jedoch, dass viele Erwachsene auch heute noch dem Zauber von Spielzeugrobotern erliegen.

Viele Errungenschaften, die ehemals ins Reich der Träume gehörten, sind inzwischen Wirklichkeit geworden und im Verlauf dieses Jahrhunderts werden zahlreiche hinzukommen. Damals, als wir noch einen Bleistift in ein Etui steckten, um einen Walkie-Talkie des Welt-raumzeitalters zu konstruieren, ahnten wir nicht, dass man schon bald an jeder Straßen-ecke Kinder mit Mobiltelefonen sehen würde. Und damals gab es keinen Zweifel, dass Roboter niemals weiche oder flüssige Bewegungen ausführen könnten. Heutzutage halten sich die Leute anstelle von echten Tieren schon Roboterhaustiere oder „tanzende" Roboter, die bewusst für den Hausgebrauch hergestellt werden. Auch wenn die Spielzeugroboter uns heute technisch anspruchslos erscheinen, so macht in einer Hightech-Welt, die immer schneller an Glanz verliert, gerade das ihre enorme Beliebtheit aus.

Spielzeug, das im 20. Jahrhundert, dem Zeitalter von Massenproduktion und Massen-konsum, weggeworfen wurde, konnte ich retten und in meine Sammlung aufneh-men. Ich hoffe, dass nachfolgende Generationen, denen es niemals vergönnt war, einen dieser Spielzeugroboter tatsächlich in der Hand zu halten, in meinen sie-ben Museen oder in diesem Buch trotzdem die Wärme spüren, die sie ausstrah-len. Ich bin der festen Überzeugung, dass sich der geheimnisvolle Charme dieser Roboter im Geiste des 21. Jahrhunderts noch sehr viel stärker ausbreiten wird.

Teruhisa Kitahara

Ma passion pour les robots
Teruhisa Kitahara

Dans mon enfance (du milieu des années 1950 au début des années 1960), je voulais faire croire que j'étais un robot, un astronaute, une fusée : je vivais dans un monde de films, de BD et d'histoires de science-fiction regorgeant de « boum ! » et de « clank ! ». A l'époque de l'aventure spatiale et du programme Apollo, les boutiques de jouets étaient pleines de robots et de fusées, et la science alimentait les rêves des hommes.

Je passais des journées entières à jouer avec mes nouveaux jouets à l'effigie des derniers héros de films ou de séries télé. Quand un jouet se cassait, je l'ouvrais et je l'inspectais avec l'attention d'un chirurgien, scrutant engrenages, ressorts, bobines et compteurs. Quelques années après, une fois ces jouets devenus des épaves disloquées, je les jetais.

Je me souviens du choc que j'ai éprouvé, quinze ans plus tard, en redécouvrant ces robots. Ce fut un étrange sentiment d'excitation, plus fort qu'une nostalgie d'adulte, pour des objets qui avaient stimulé mon imagination enfantine.

L'un des attraits de la redécouverte des robots pour enfants réside dans la couleur, la texture du matériau, et la minutie des détails qui les caractérisent. La majorité d'entre eux, fabriqués dans les années 1950 et 1960, portent la mention « Made in Japan ». Ils étaient, pour la plupart, exportés aux Etats-Unis, et leur emballage, conçu en anglais, paraît aussi moderne qu'autrefois. Je suis toujours ébahi à l'idée de fabricants de jouets consacrant tant de passion et de moyens techniques à créer une mini-salle de contrôle informatique pour une station spatiale ou un robot télécommandé.

Cette approche créative confère à ces robots, bien au-delà de leur nature de simples jouets, une solidité, une présence physique qui résiste à leur transposition graphique ou photographique. C'est sans doute la raison pour laquelle mes robots sont si souvent utilisés dans des publicités ou des couvertures de disques.

Autre aspect intéressant des robots : les multiples variantes suivant la période considérée. Par exemple aux débuts de la télévision, des robots en forme de télé ont fait leur apparition. Quand Apollo a atterri sur la Lune, on a fabriqué des robots de l'espace, des astronautes, des stations spatiales... un personnage populaire comme Robby, le robot de « Forbidden Planet » (planète interdite), un film de science-fiction à grand succès de 1956, a été fabriqué à plus de 300 exemplaires.

Alors que je venais juste de commencer ma collection, fasciné par l'impressionnante vitalité de ce genre, je suis tombé, dans une librairie étrangère de Tokyo, sur un livre français intitulé « Robots ». Et j'ai été extrêmement surpris de découvrir qu'il existait des personnes, dans des pays éloignés, qui partageaient mon intérêt. Aujourd'hui, les collectionneurs sont pour la plupart américains et c'est aux Etats-Unis que les robots de collection atteignent des prix plus élevés. J'ai l'impression que les robots sont devenus une source de communication internationale et qu'ils exercent une attraction planétaire.

Les robots sont un très vieux rêve humain, mais si futuristes soient-ils, ces jouets nous rappellent l'ingéniosité et le savoir-faire des artisans des années 1950. Je serais incapable de dire si les hommes sont ensorcelés par les machines, mais je sais que beaucoup d'adultes sont captivés par les robots pour enfants.

Les rêves d'autrefois deviennent peu à peu les réalités d'aujourd'hui et du siècle qui commence. A l'époque où un crayon collé à l'oreille devenait un talkie-walkie de l'ère spatiale, personne n'aurait imaginé que les rues seraient un jour pleines d'enfants bavardant sur leurs téléphones mobiles. De même, il était admis que les robots ne seraient jamais capables de reproduire la souplesse de mouvements d'une démarche humaine. Aujourd'hui, les gens préfèrent les chiens-robots aux chiens réels et chacun peut acheter un « robot danseur ». Si « low tech » que nous paraissent ces humbles robots analogiques, c'est précisément cette caractéristique qui les rend populaires, dans un monde « high tech » qui perd rapidement de son lustre.

Les jouets qui auraient dû finir dans les poubelles de ce XXe siècle de « production et de consommation de masse » ont survécu et sont venus accroître ma collection. Je leur ai donné une seconde chance. J'espère que les générations montantes, qui n'ont jamais eu le plaisir de manipuler ces robots, ressentiront la chaleur qu'ils dégagent dans mes sept musées ou dans ce livre. Je crois aussi que le charme mystérieux des robots s'épanouira pleinement dans l'esprit du XXIe siècle.

Teruhisa Kitahara

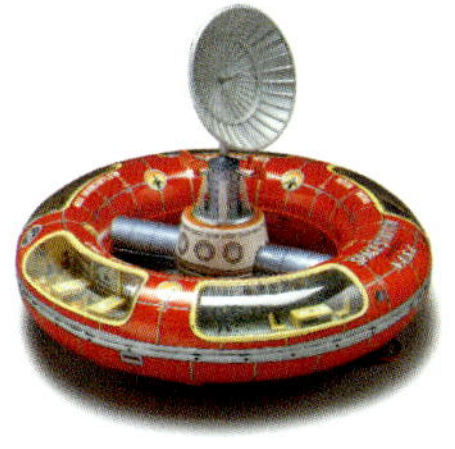

Robots and Spaceships

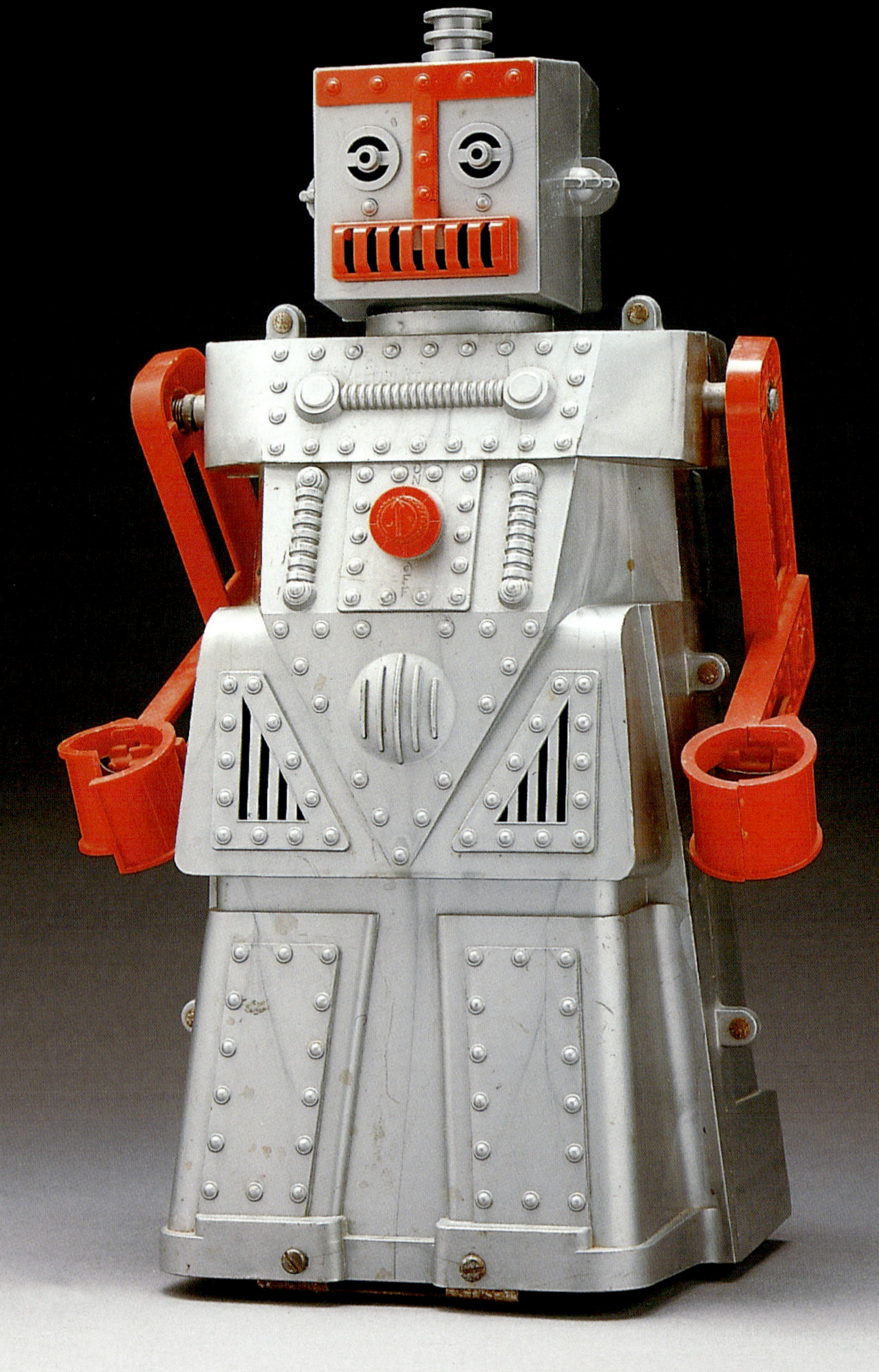

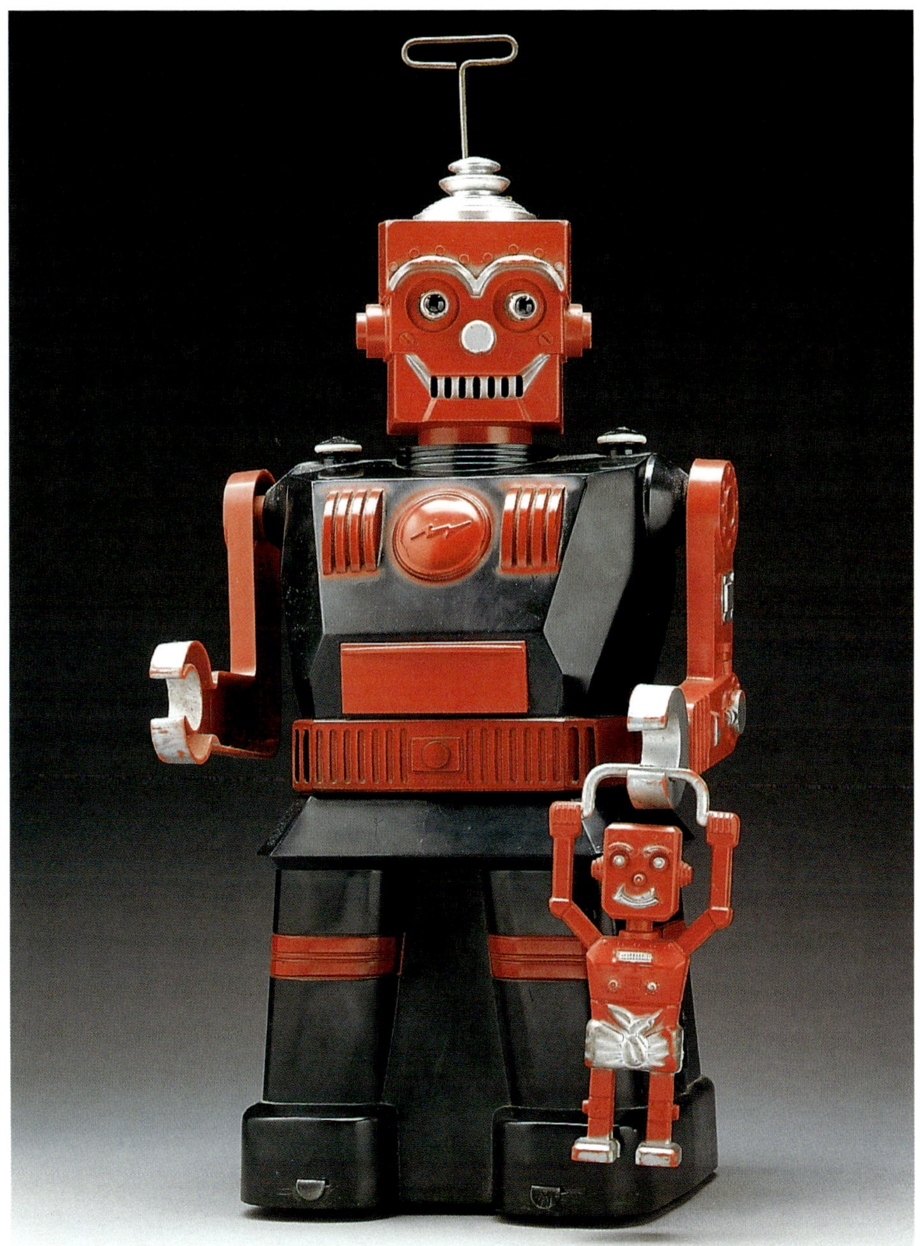

▲ 1950s, *Robot and Son*, Louis Marx, 150 x 185 x 365 mm

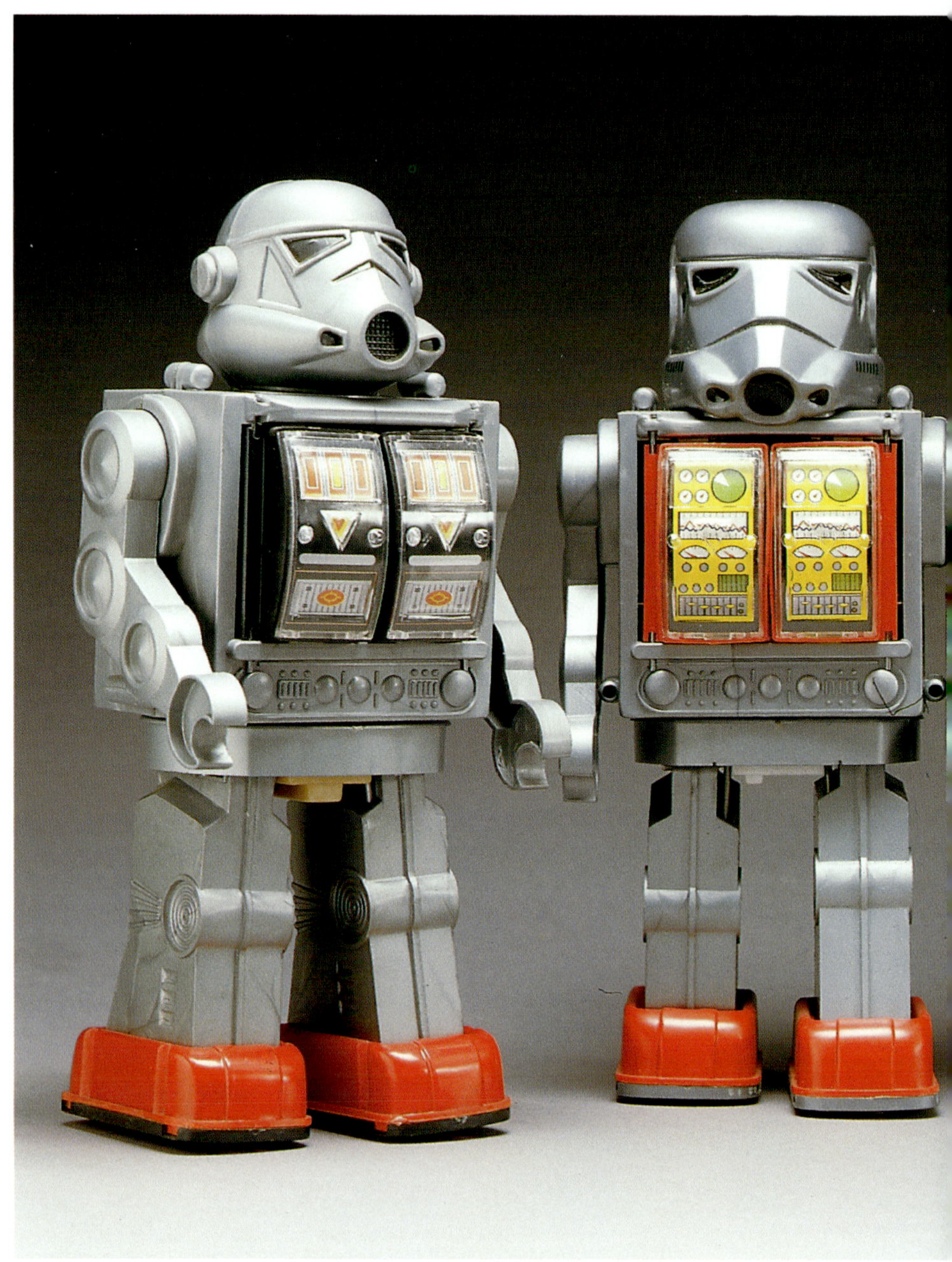

1970s, *Robots*, Hong Kong, 98 x 140 x 280 mm

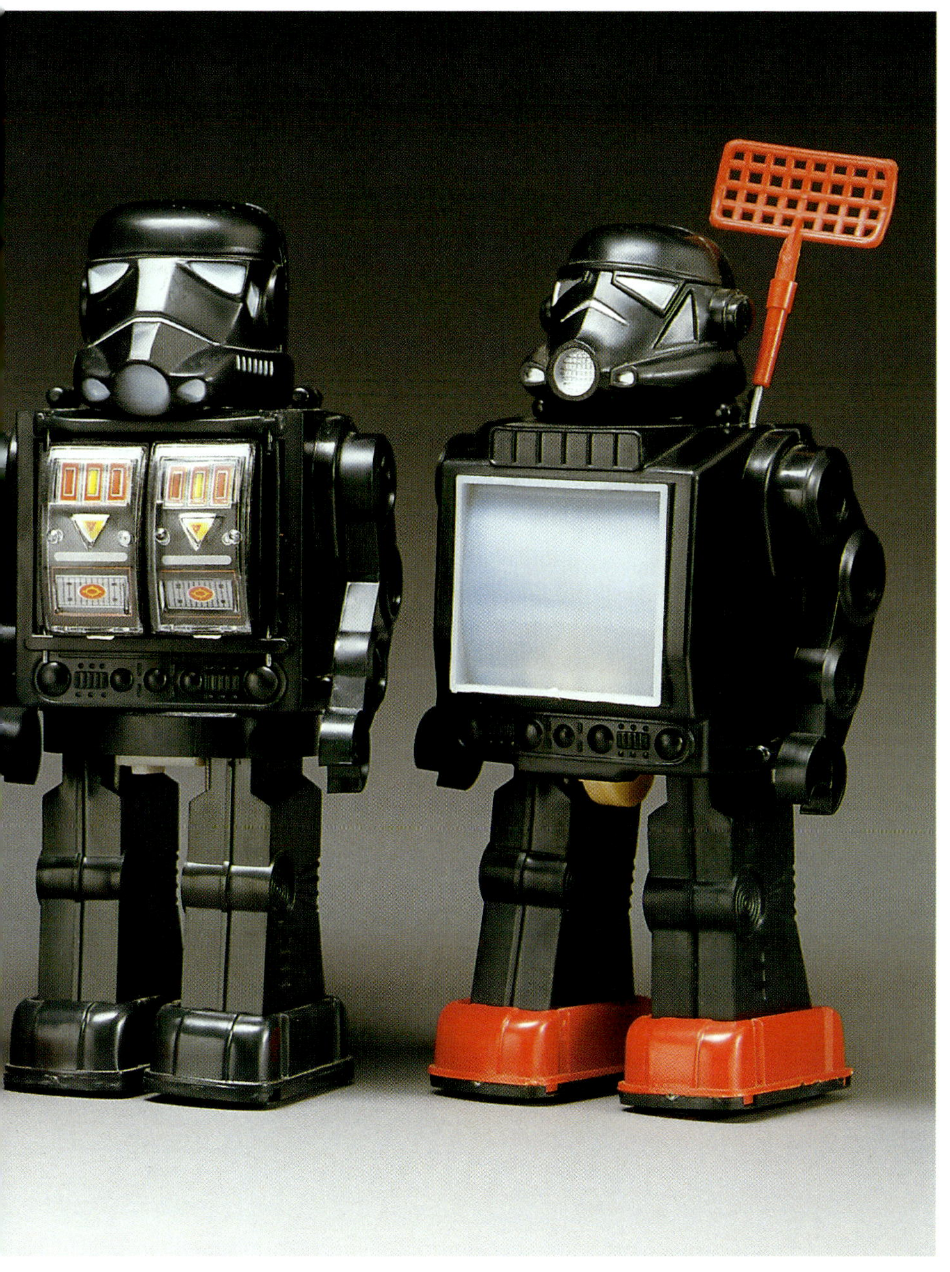

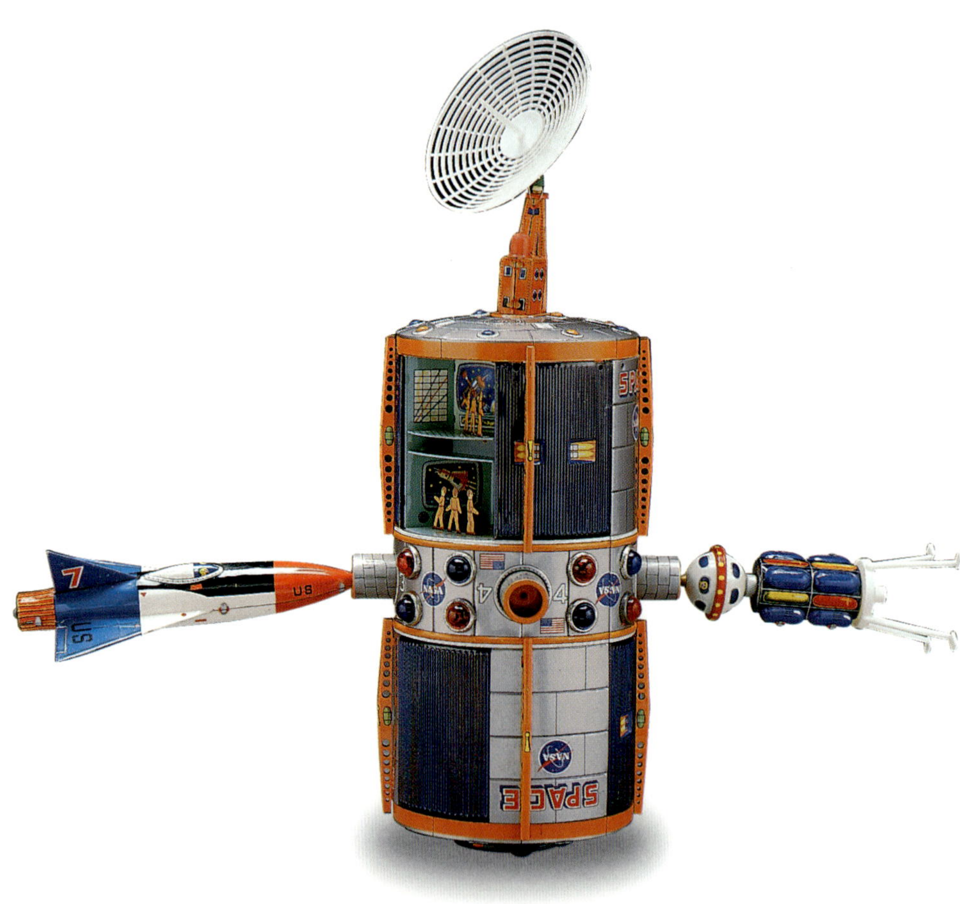

1950s, *Space Station*, Waco, 150 x 445 x 380 mm

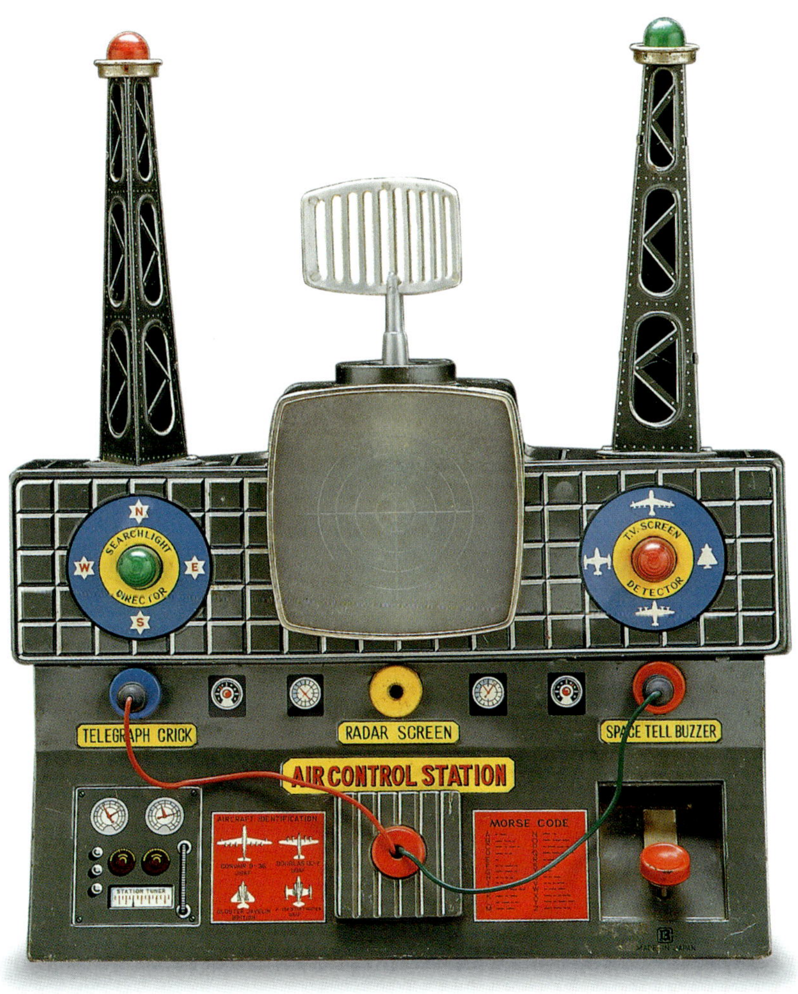

1950s, *Air Control Station*, unknown

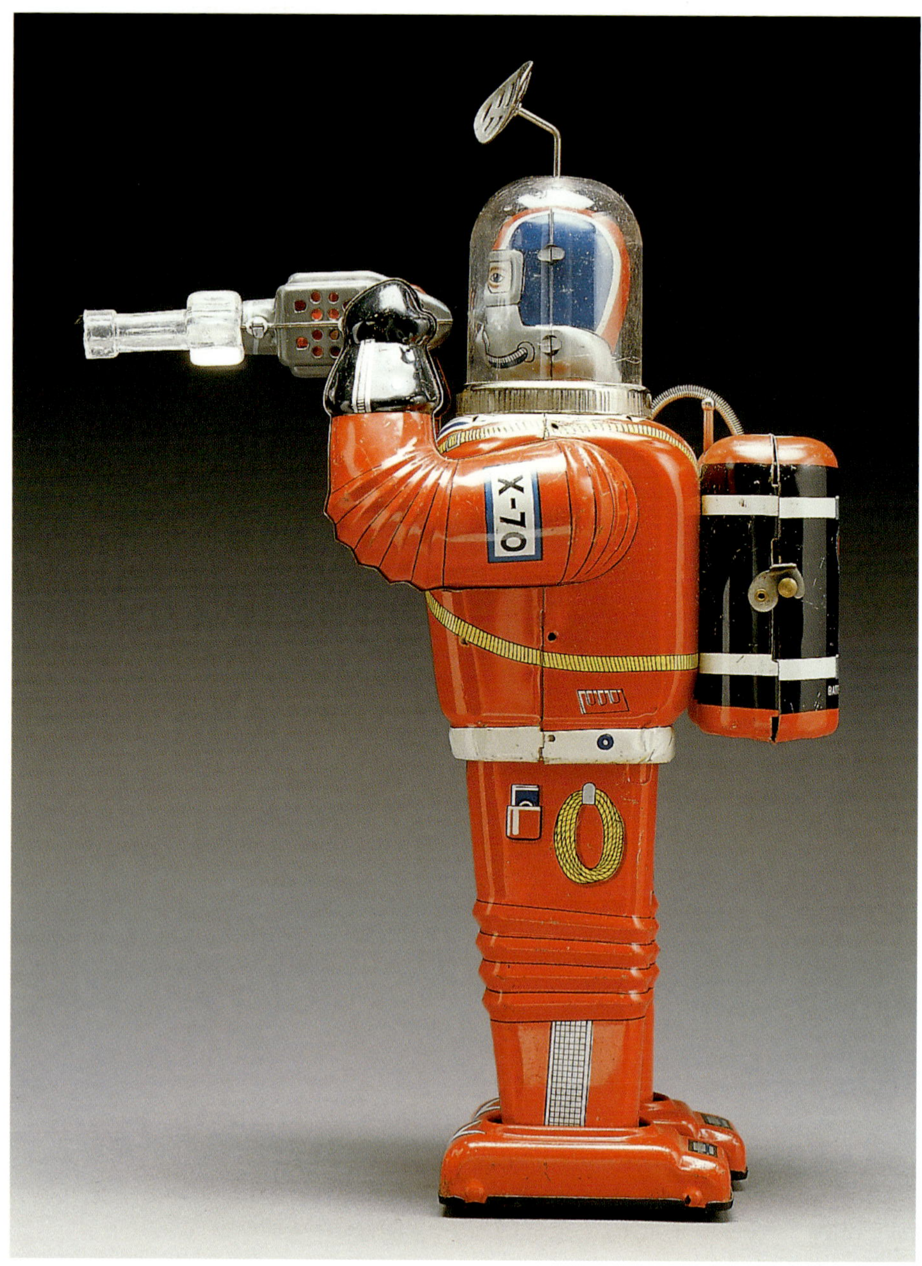

1950s, *Astronaut*, Daiya, 137 x 110 x 298 mm

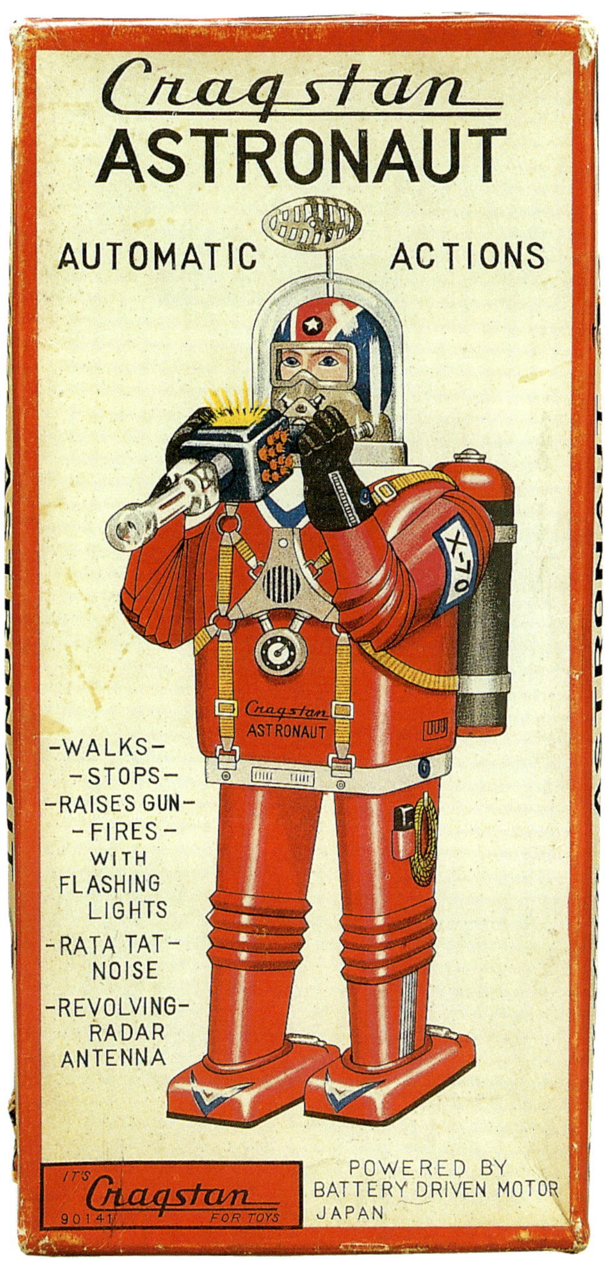
Cragstan
ASTRONAUT
AUTOMATIC ACTIONS
X-70
Cragstan
ASTRONAUT
-WALKS-
-STOPS-
-RAISES GUN-
-FIRES-
WITH
FLASHING
LIGHTS
-RATA TAT-
NOISE
-REVOLVING-
RADAR
ANTENNA
IT'S Cragstan FOR TOYS
90141
POWERED BY
BATTERY DRIVEN MOTOR
JAPAN
ASTRONAUT
ASTRONAUT

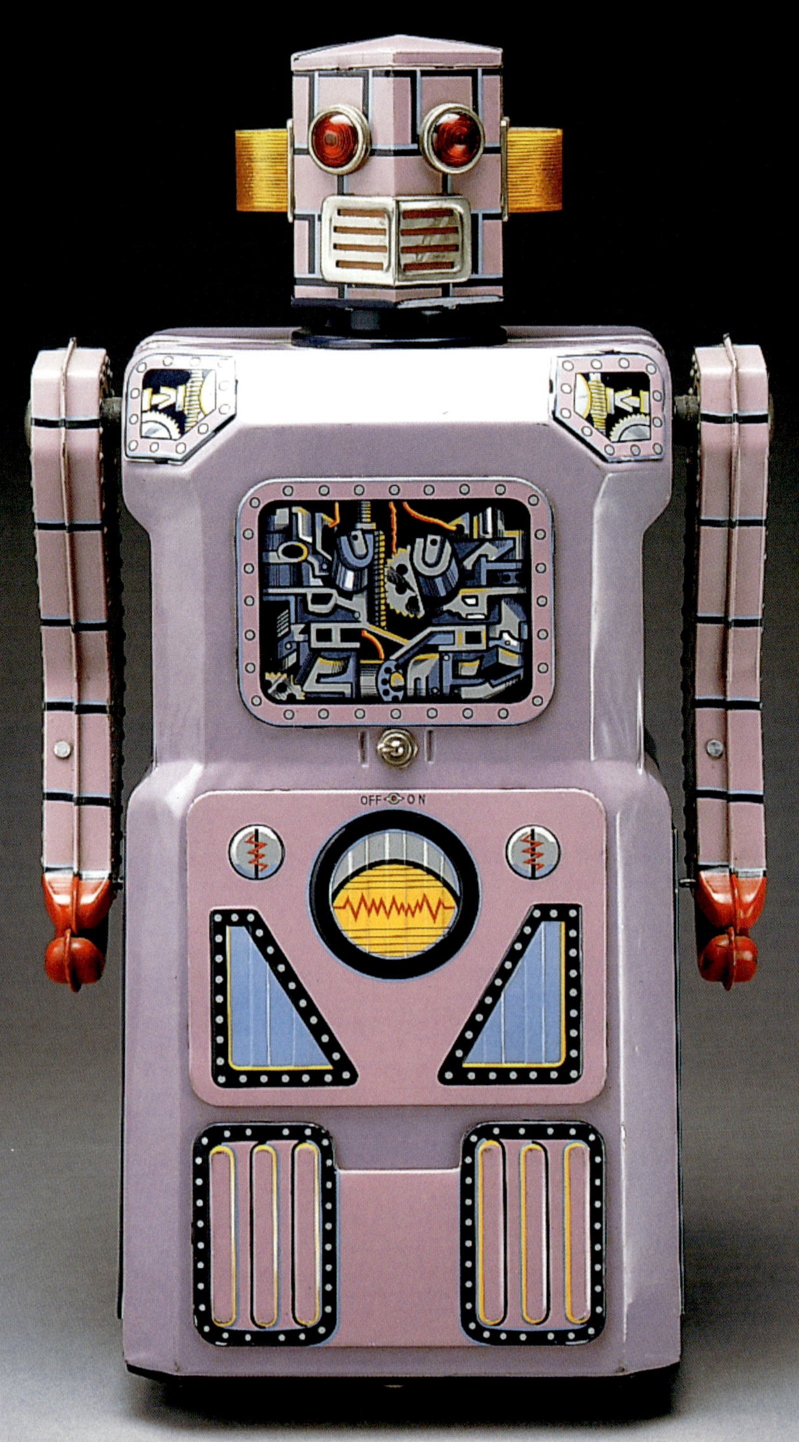

OFF ON

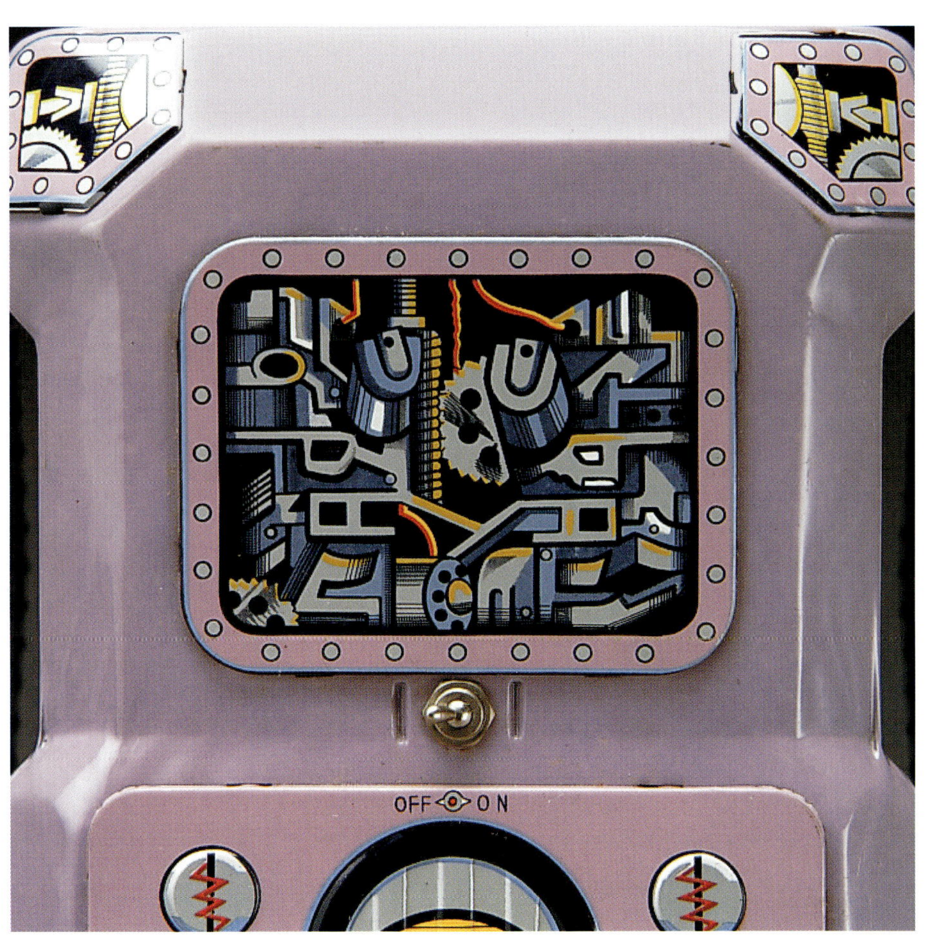

1950s, *Nonstop Robot*, Masudaya, 155 x 210 x 370 mm

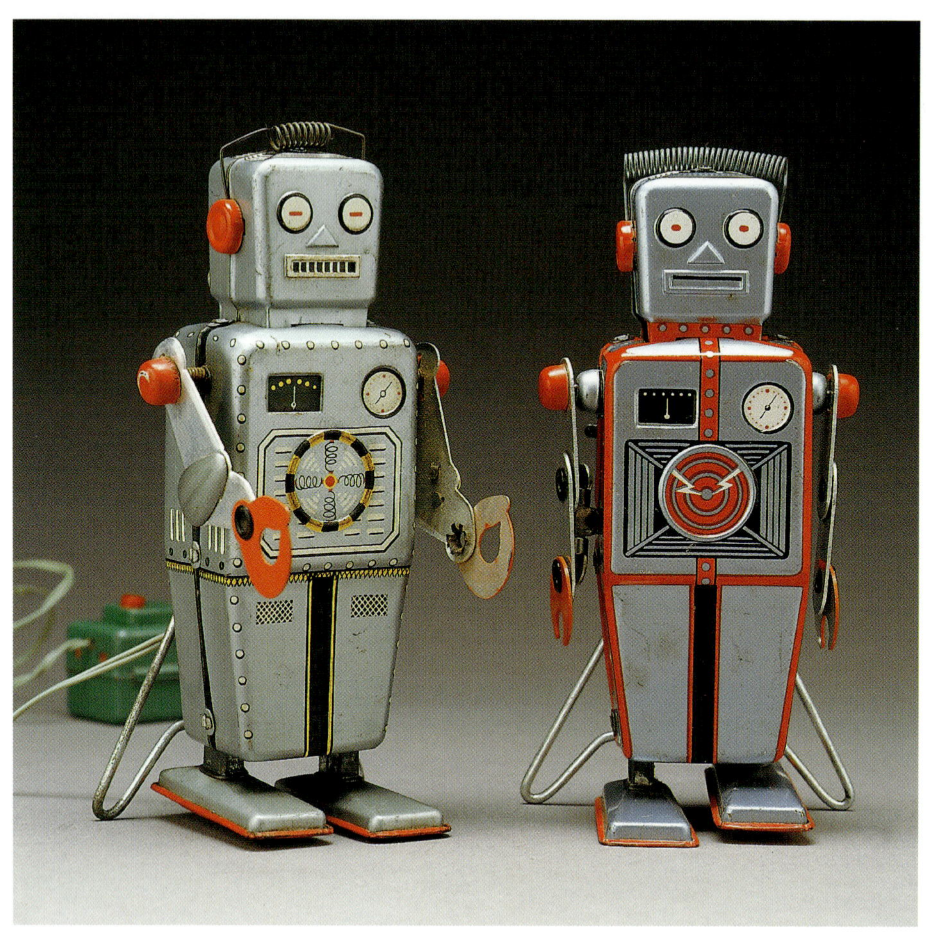

1950s, *Robots*, Yonezawa, 52 x 70 x 150 mm

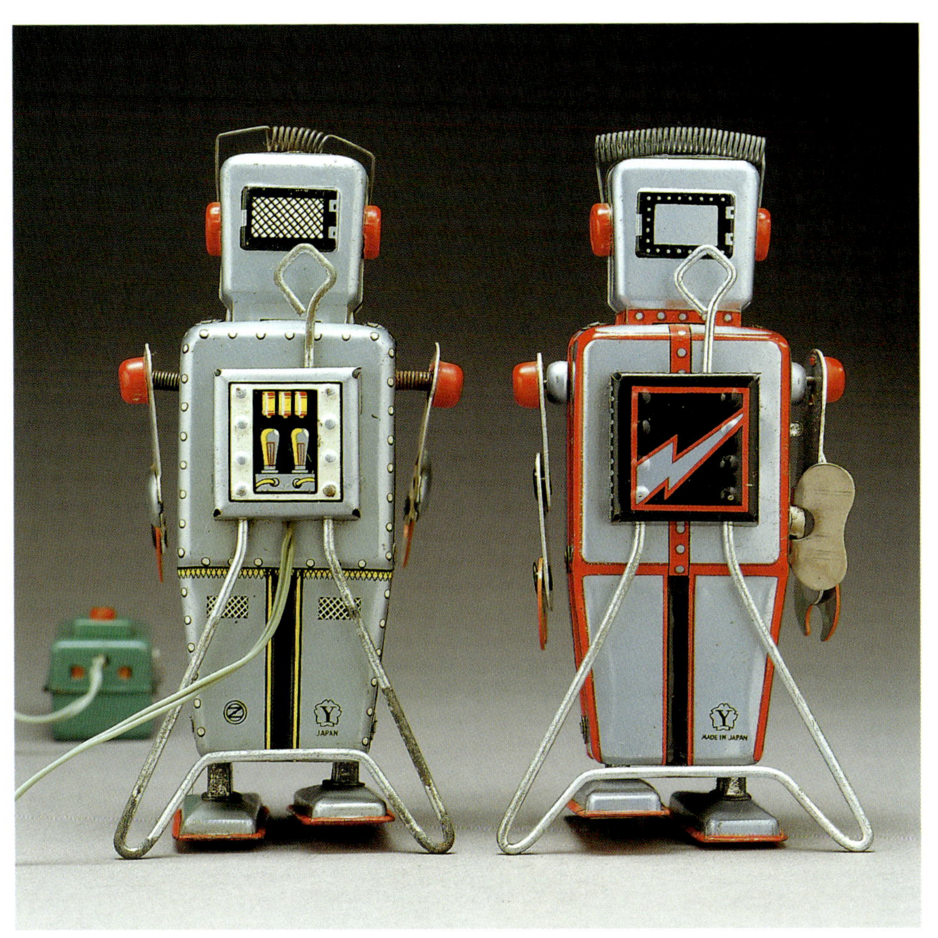

Y
JAPAN
Y
MADE IN JAPAN

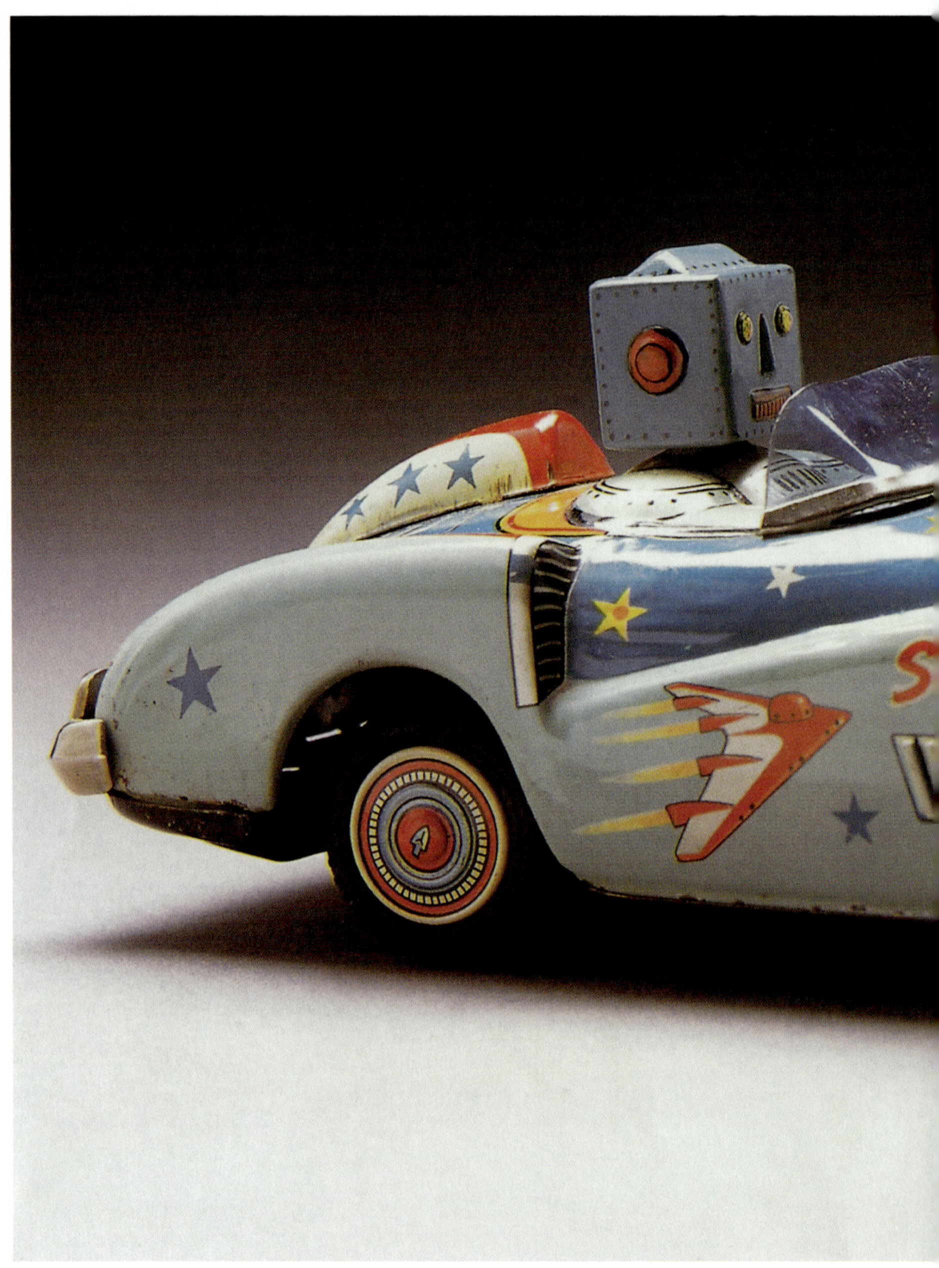

1950s, *Space Patrol Car and Box*, Ichiko, 217 x 85 x 82 mm

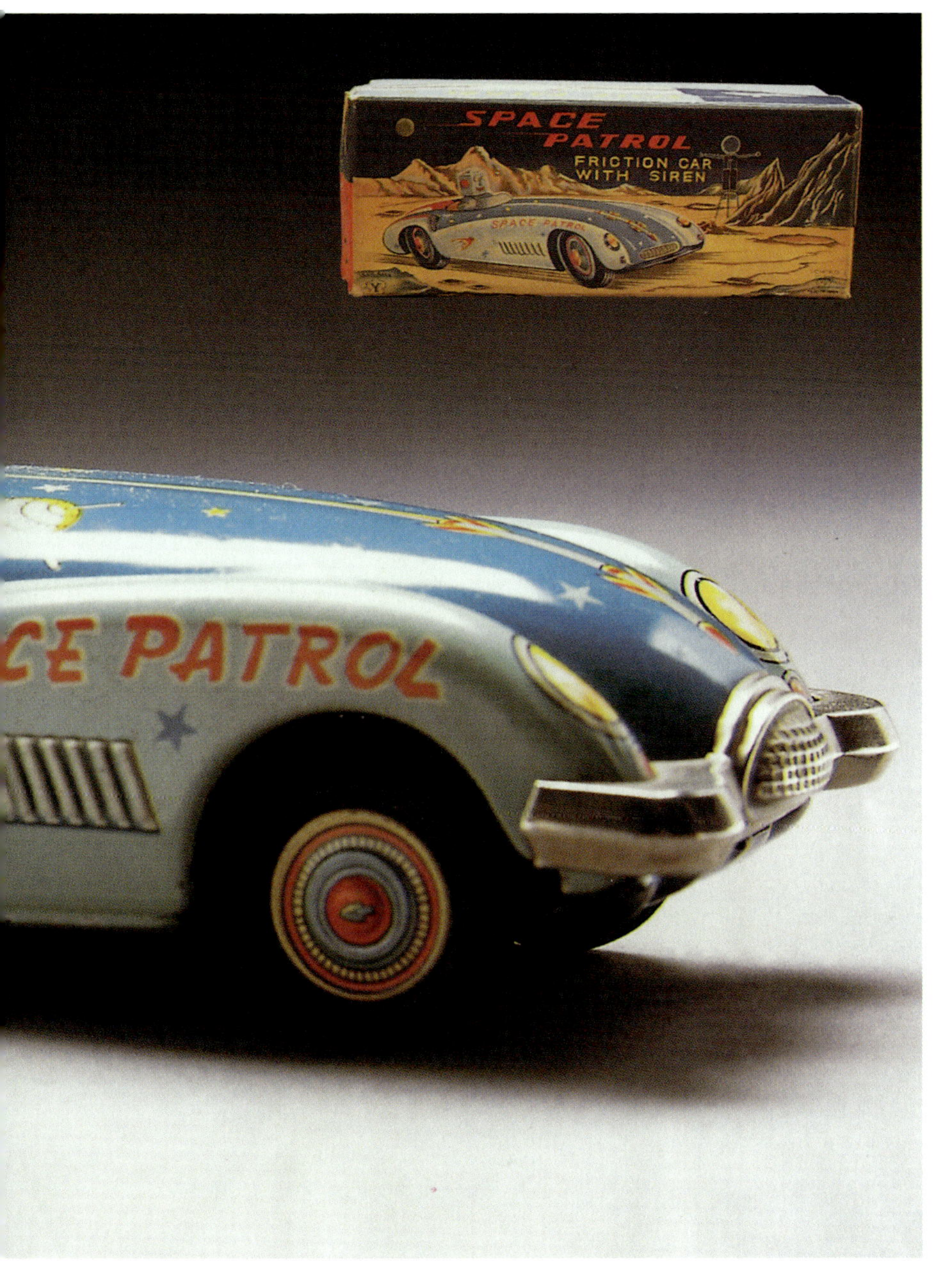

SPACE PATROL
FRICTION CAR
WITH SIREN
SPACE PATROL

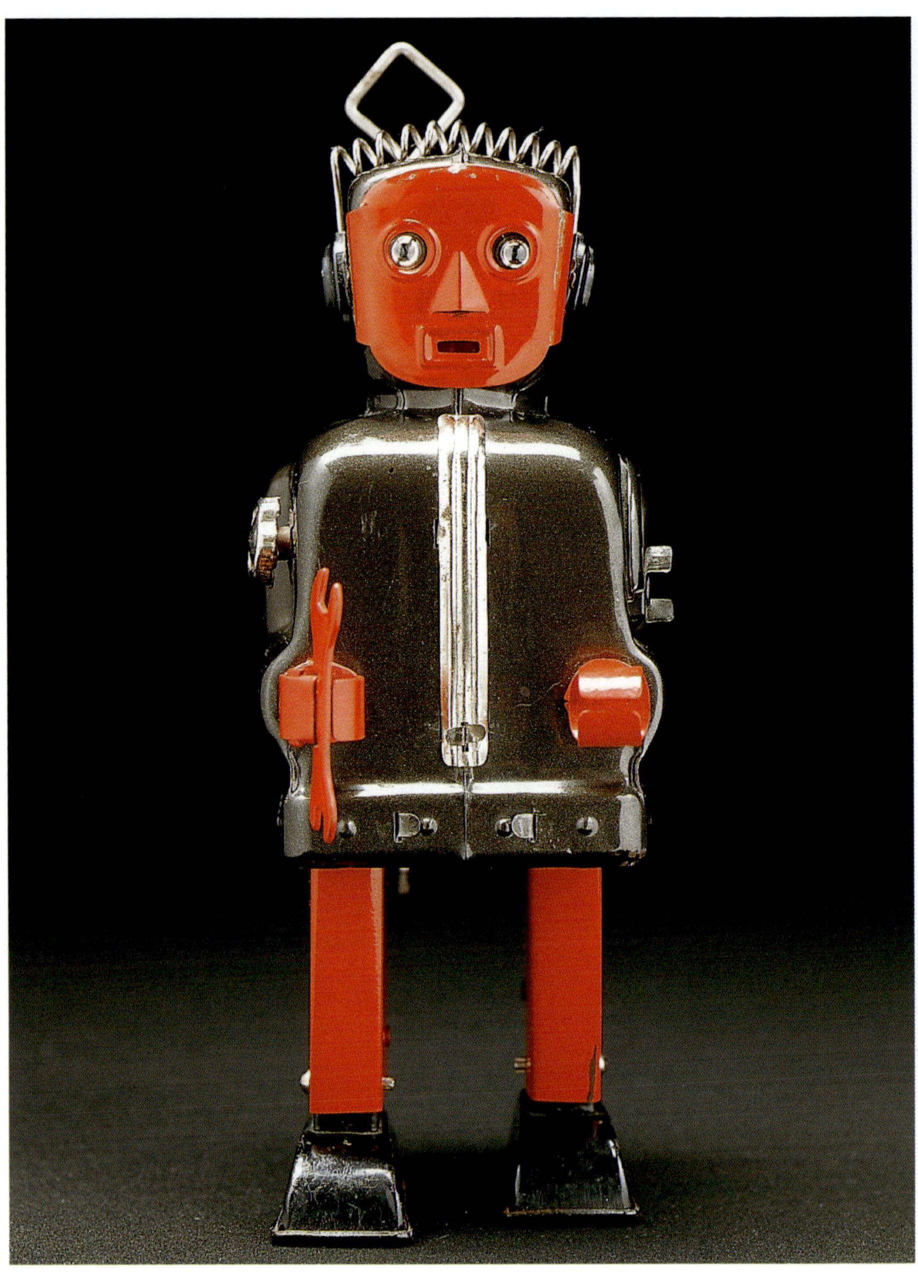

1950s, *Zoomer the Robot*, Nomura, 100 x 70 x 190 mm

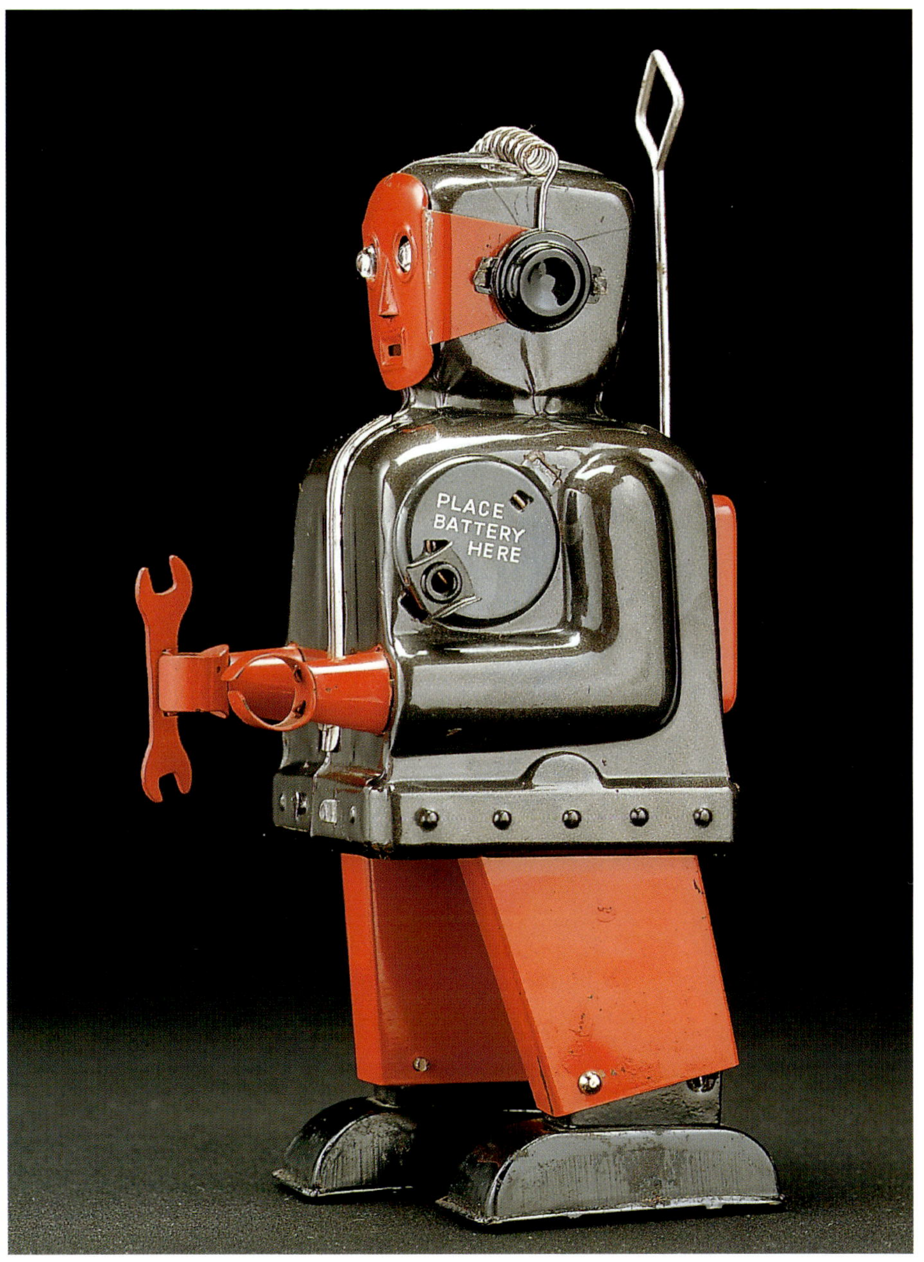

PLACE
BATTERY
HERE

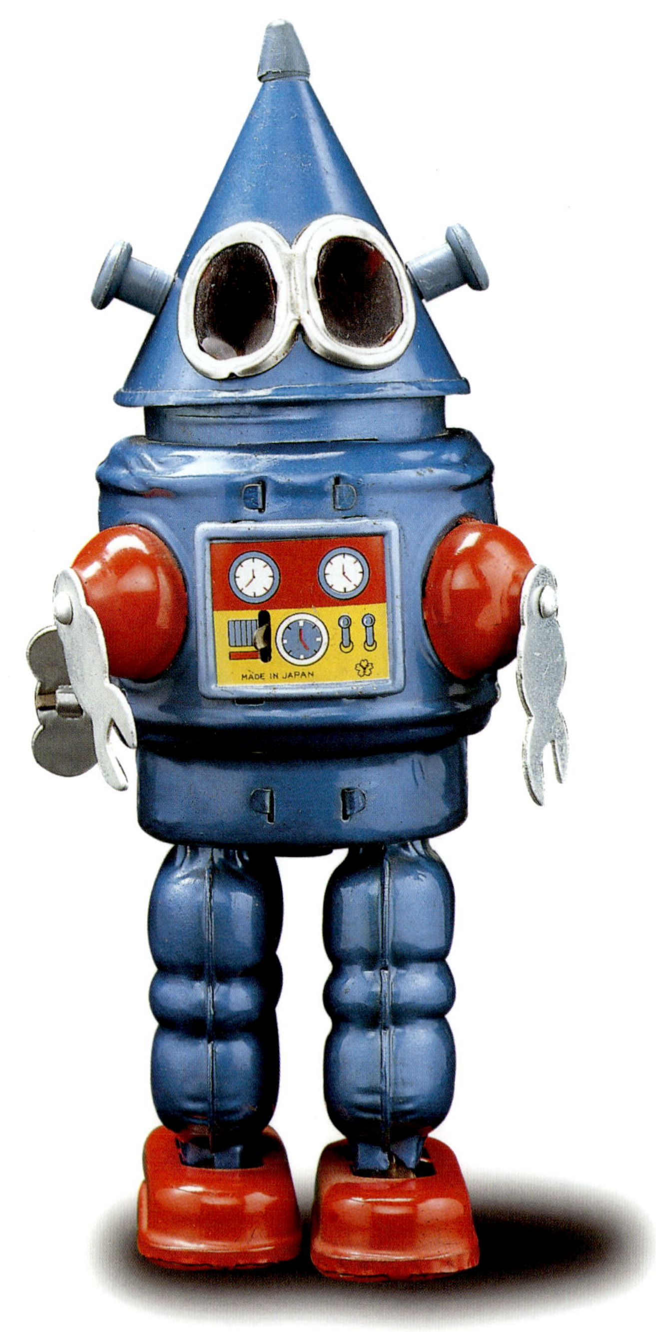

1950s, *Robot*, Yonezawa, 70 x 85 x 215 mm

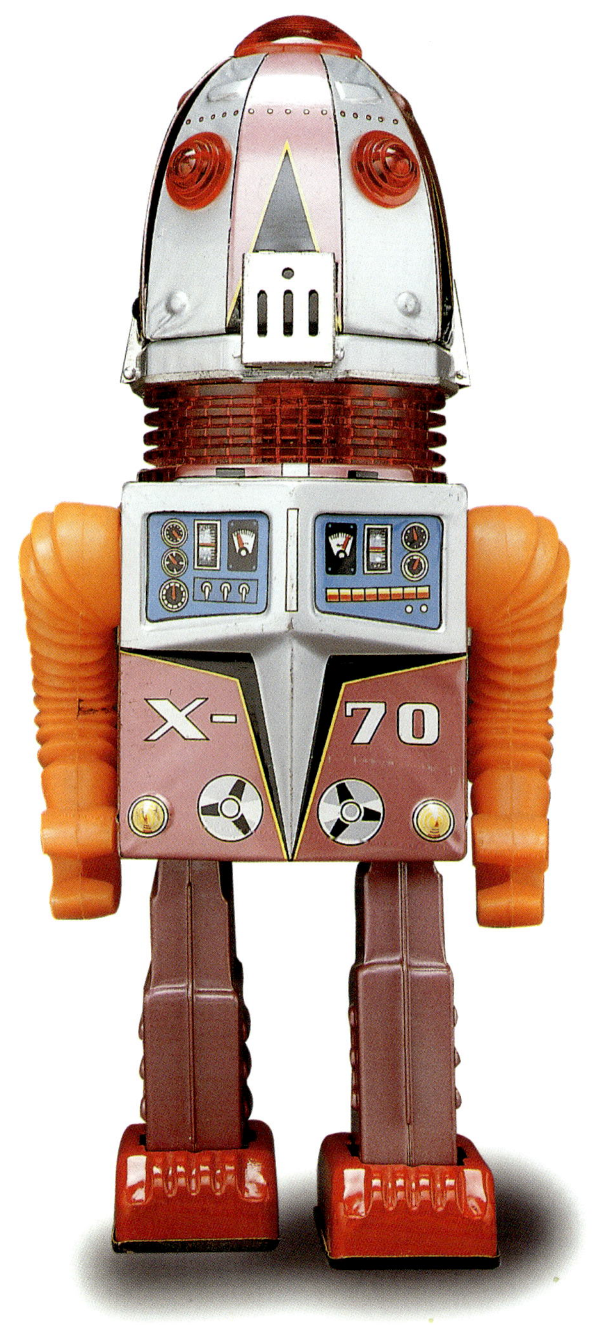

1960s, *X-70 Robot*, unknown, 90 x 130 x 305 mm

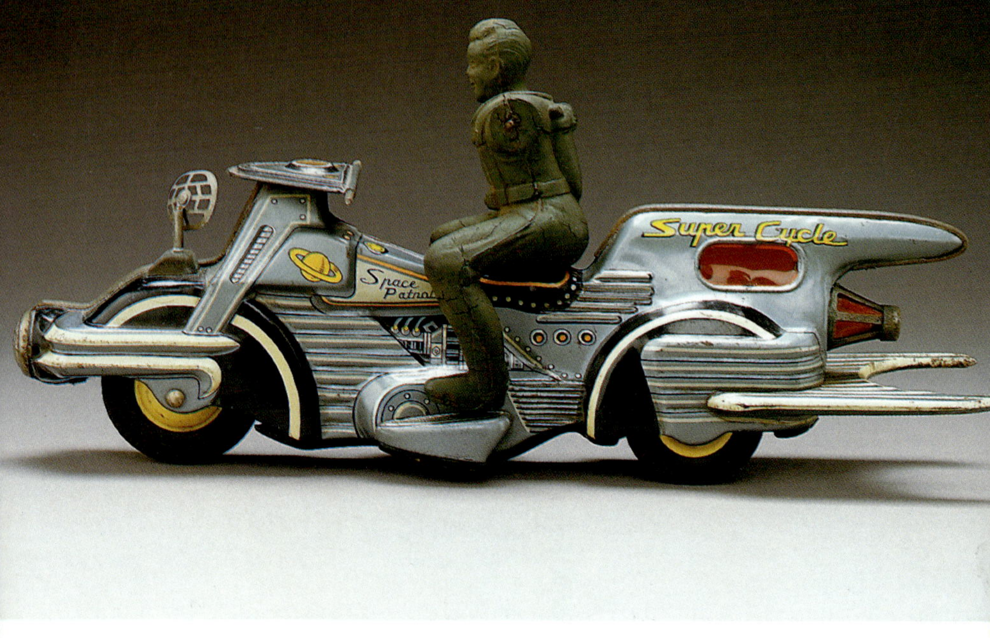

1950s, *Super-Bike*, Bandai, 305 x 70 x 138 mm

1950s, *Space Tank*, unknown, 203 x 118 x 105 mm

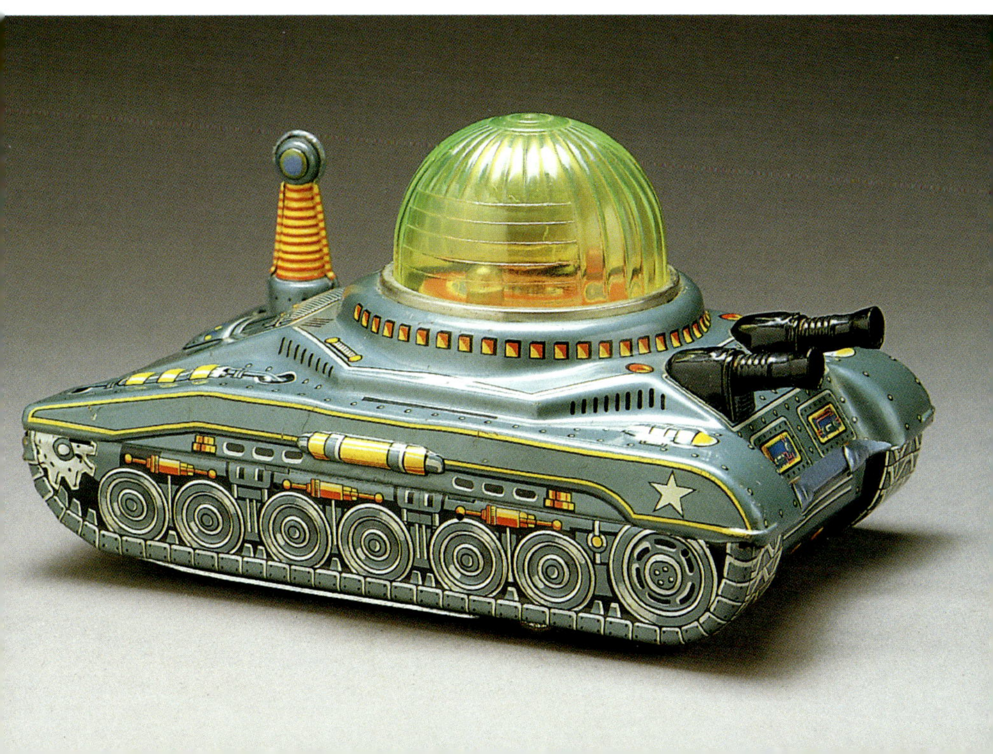

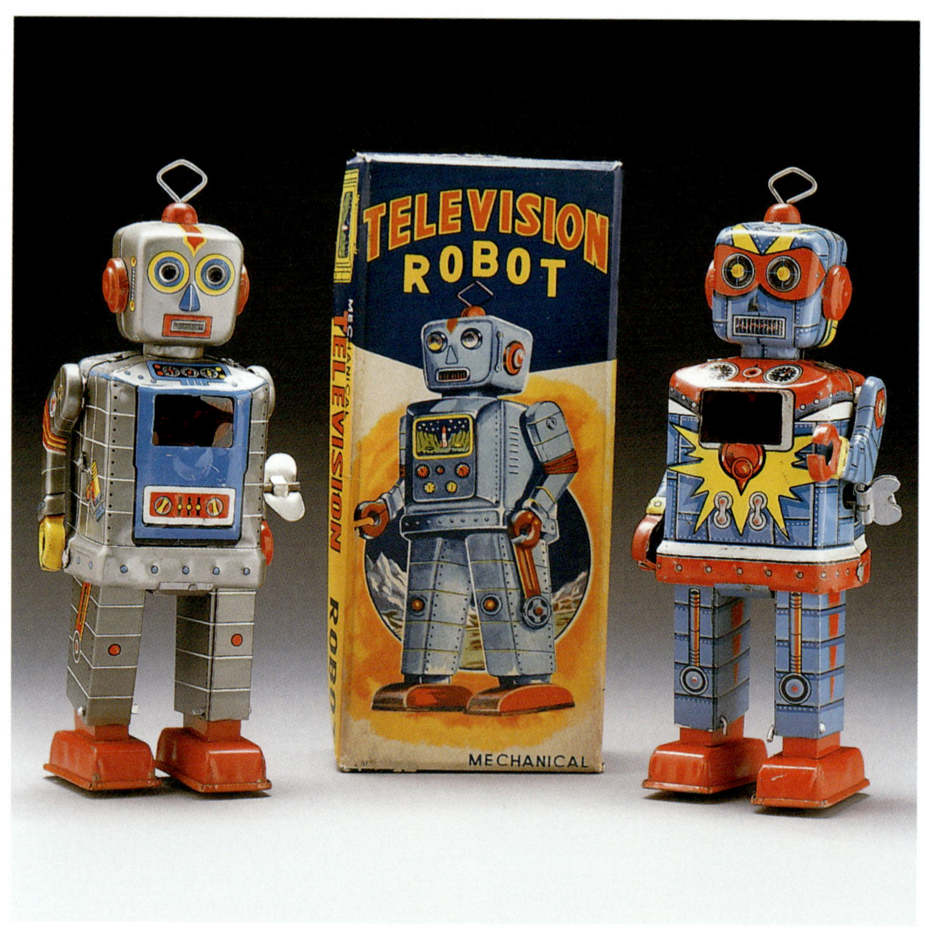

1950s, *Television Robots and Case*, Sankei, 65 x 80 x 198 mm

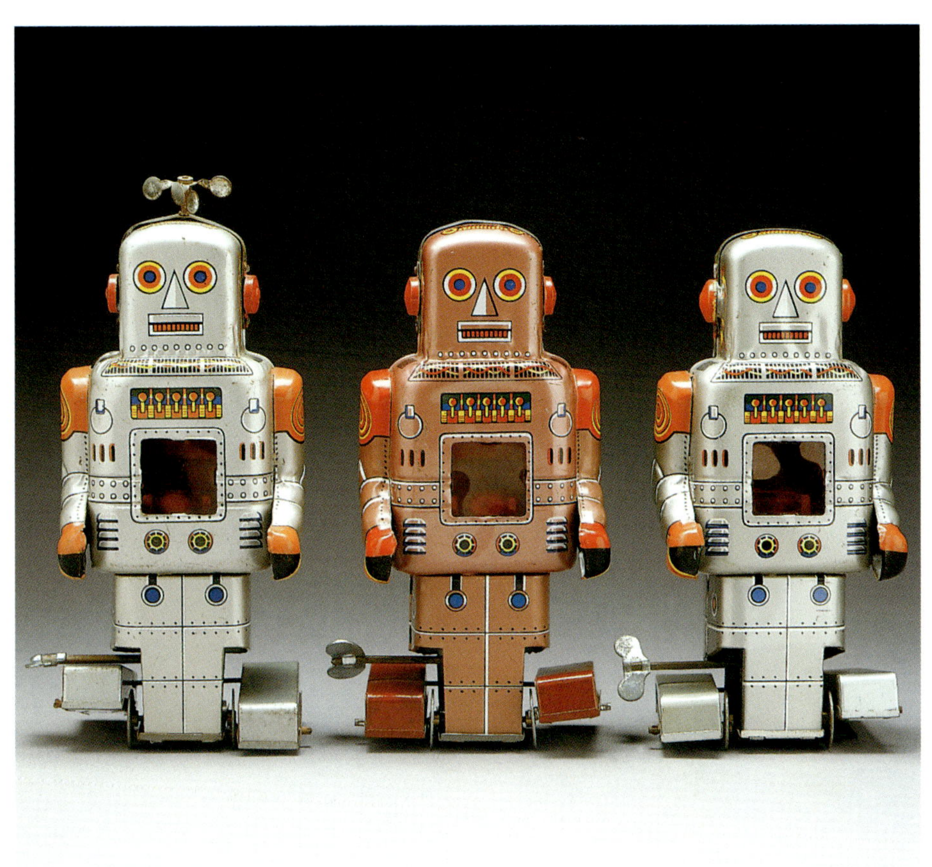

1950s, *Sparky Robots*, SY Toys, 56 x 80 x 190, 56 x 180 x 175, 56 x 80 x 175 mm

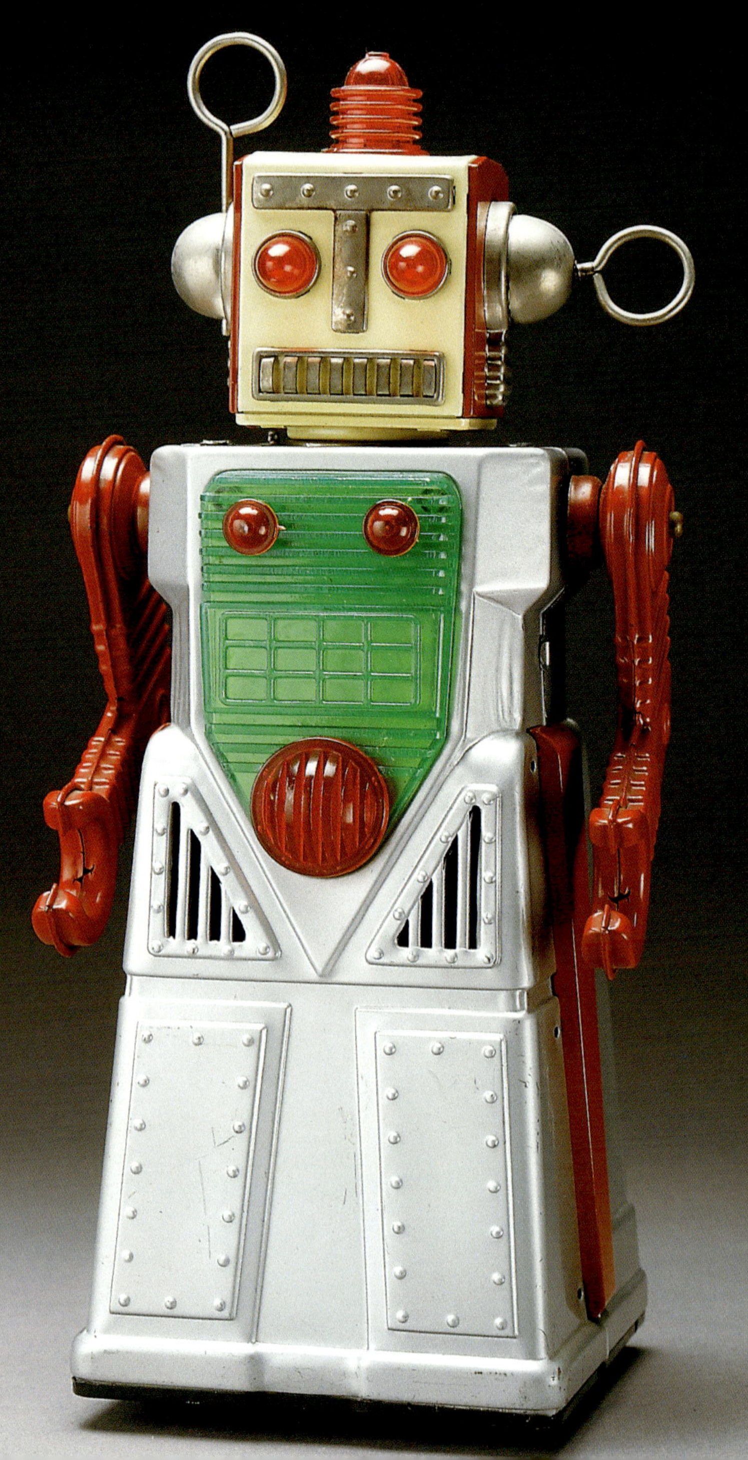

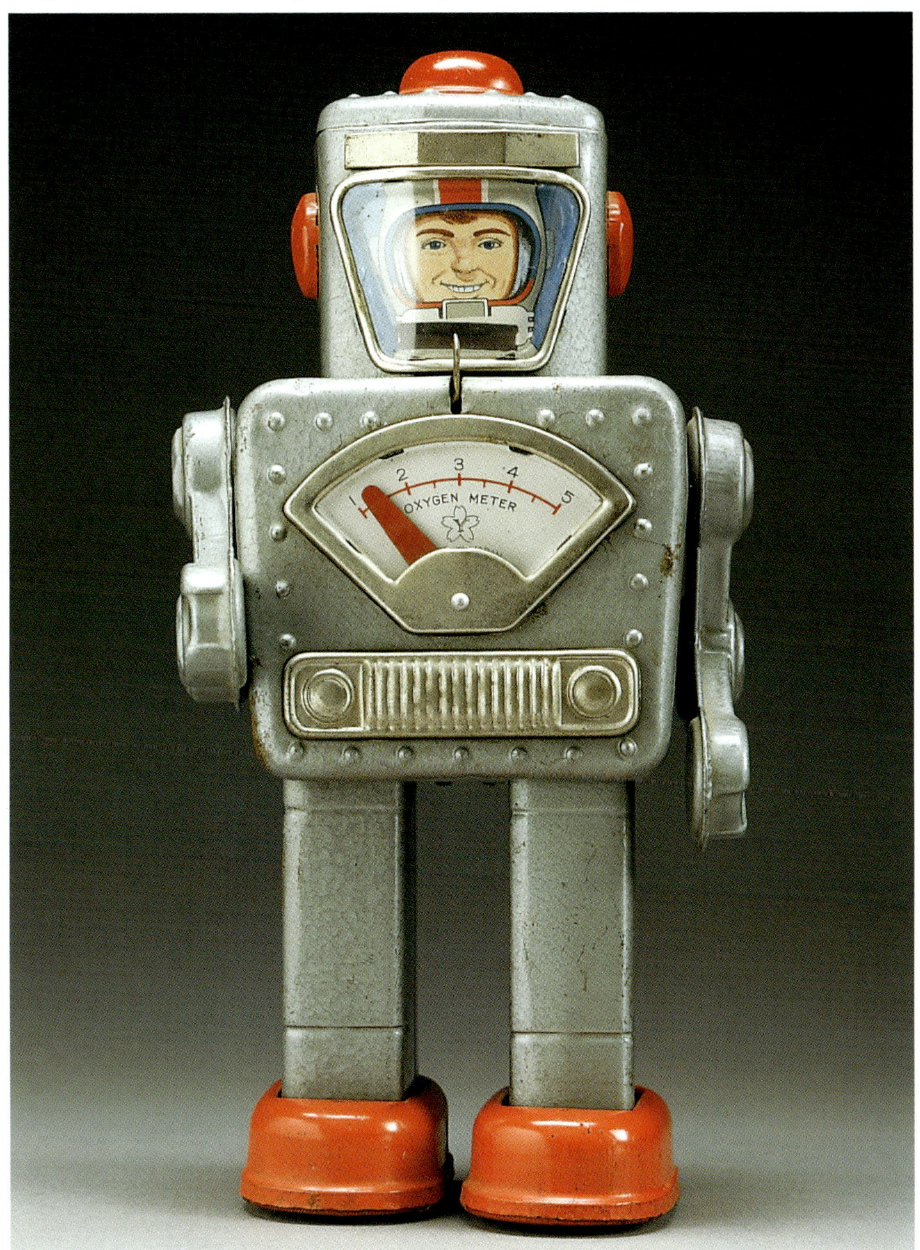

▲ 1950s, *Space Explorer*, Yonezawa, 70 x 120 x 230 mm

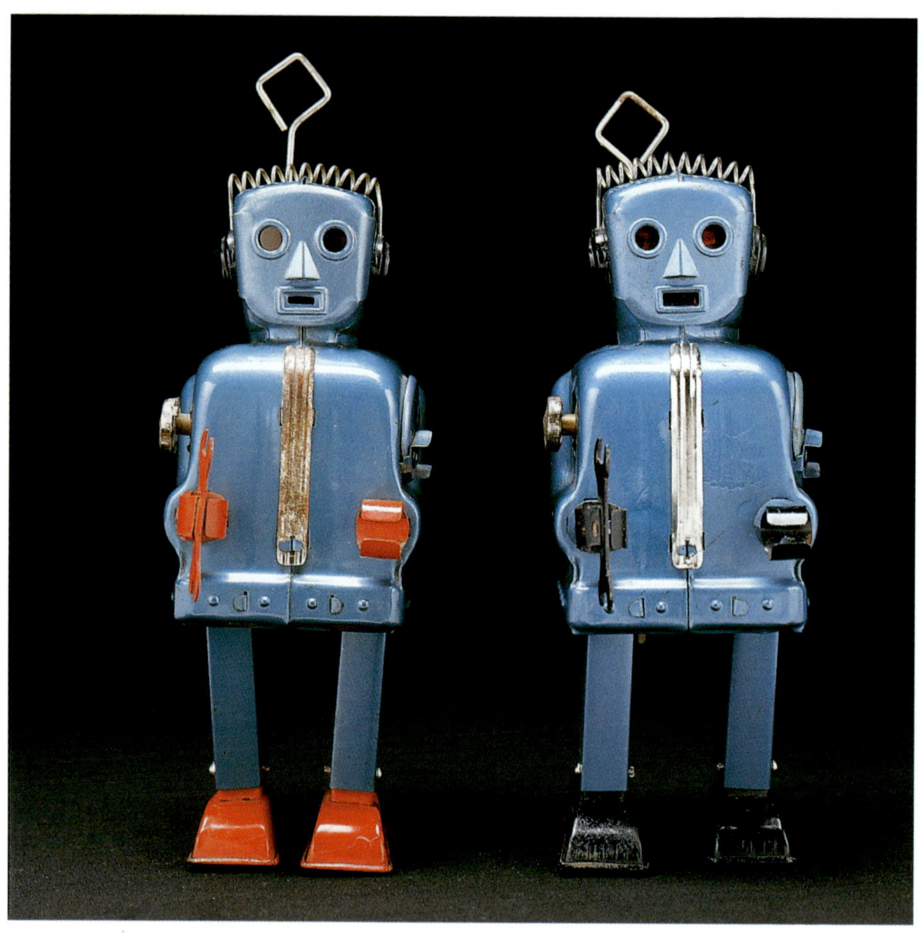

▲ 1950s, *Zoomer the Robot*, Nomura, 100 x 70 x 190 mm

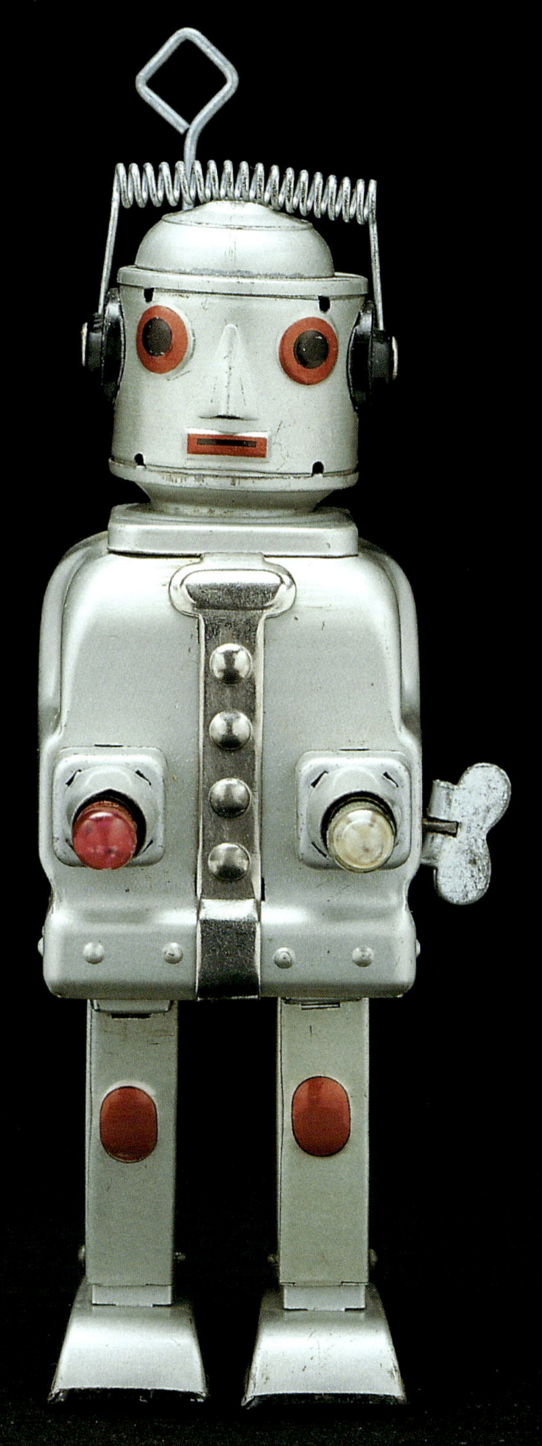

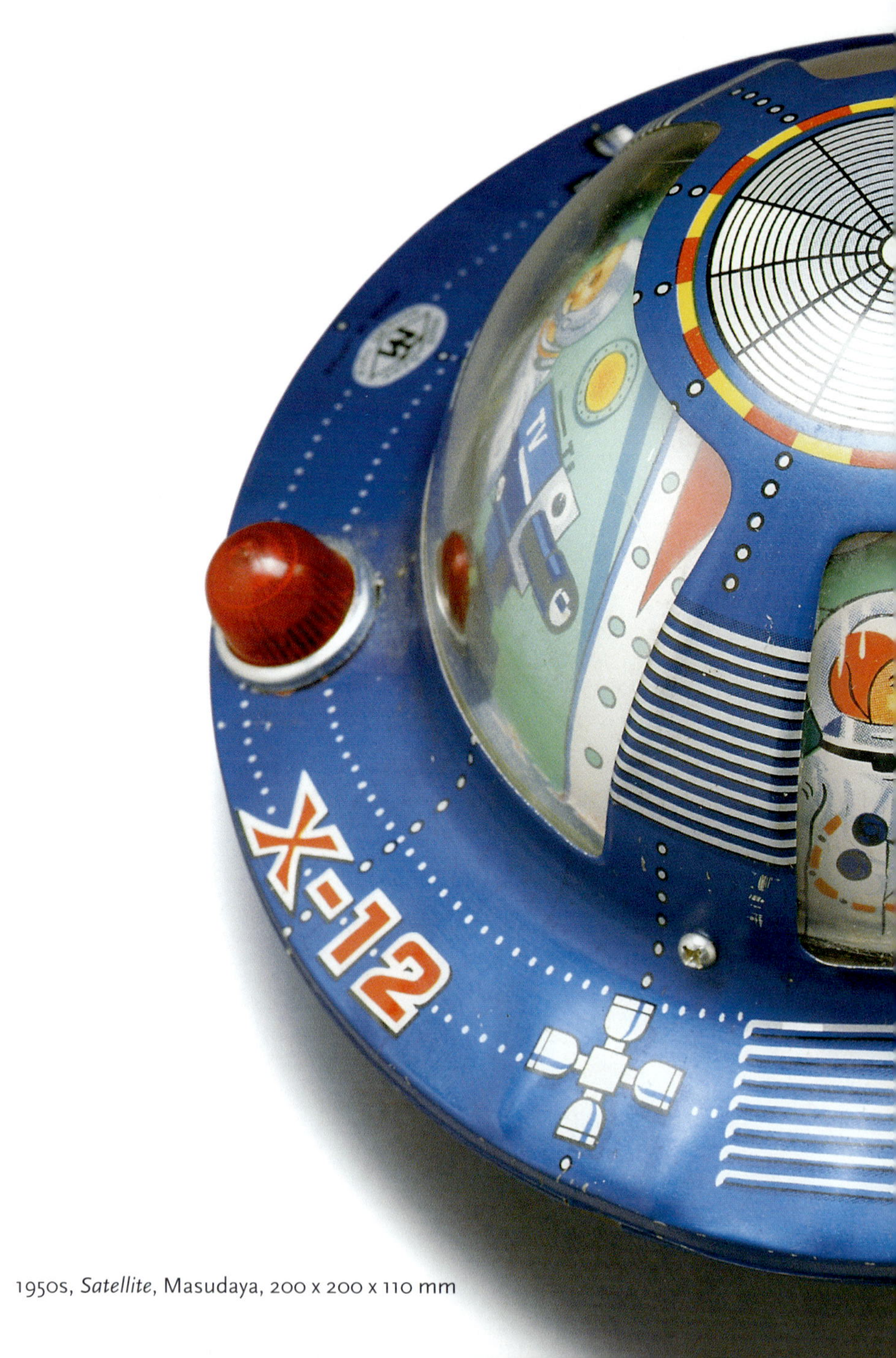

1950s, *Satellite*, Masudaya, 200 x 200 x 110 mm

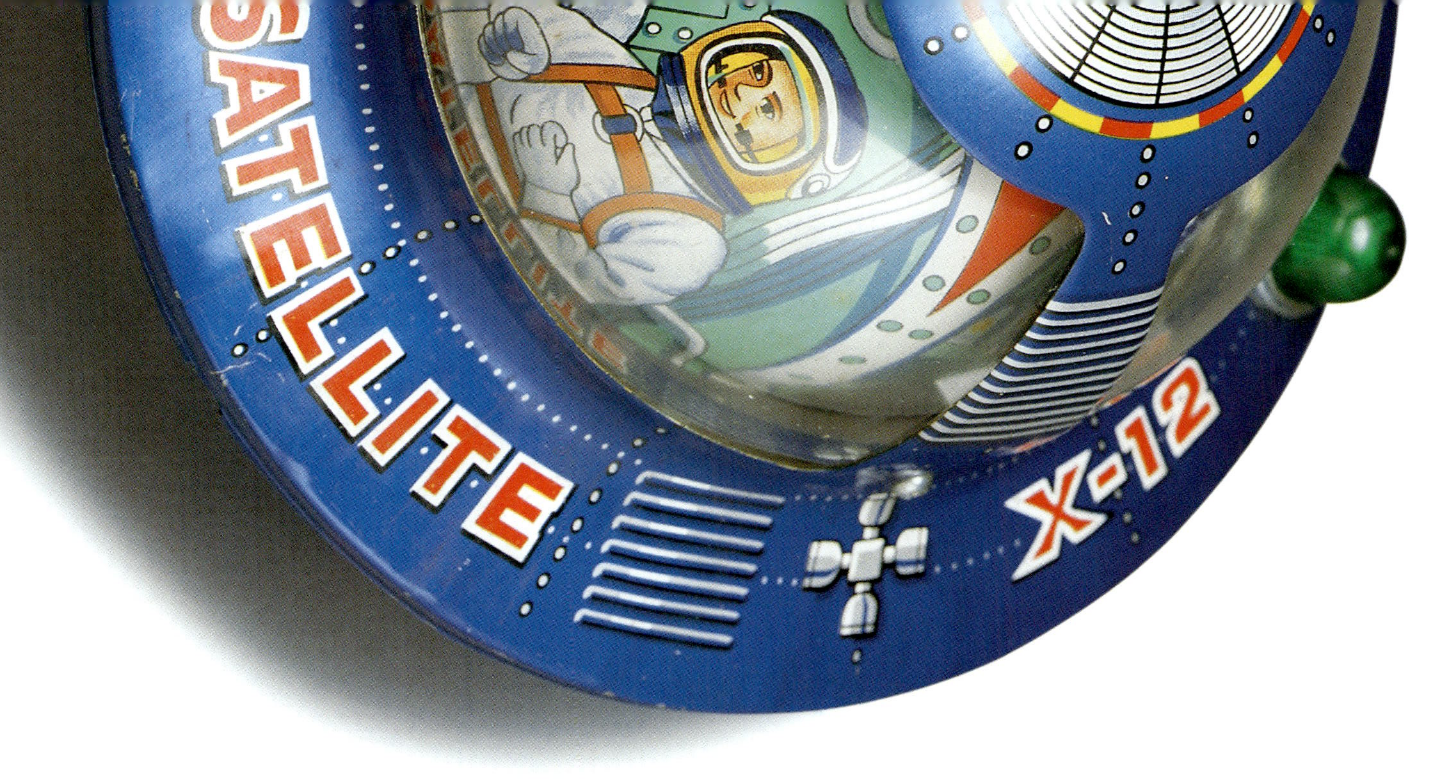

SATELLITE
X-12

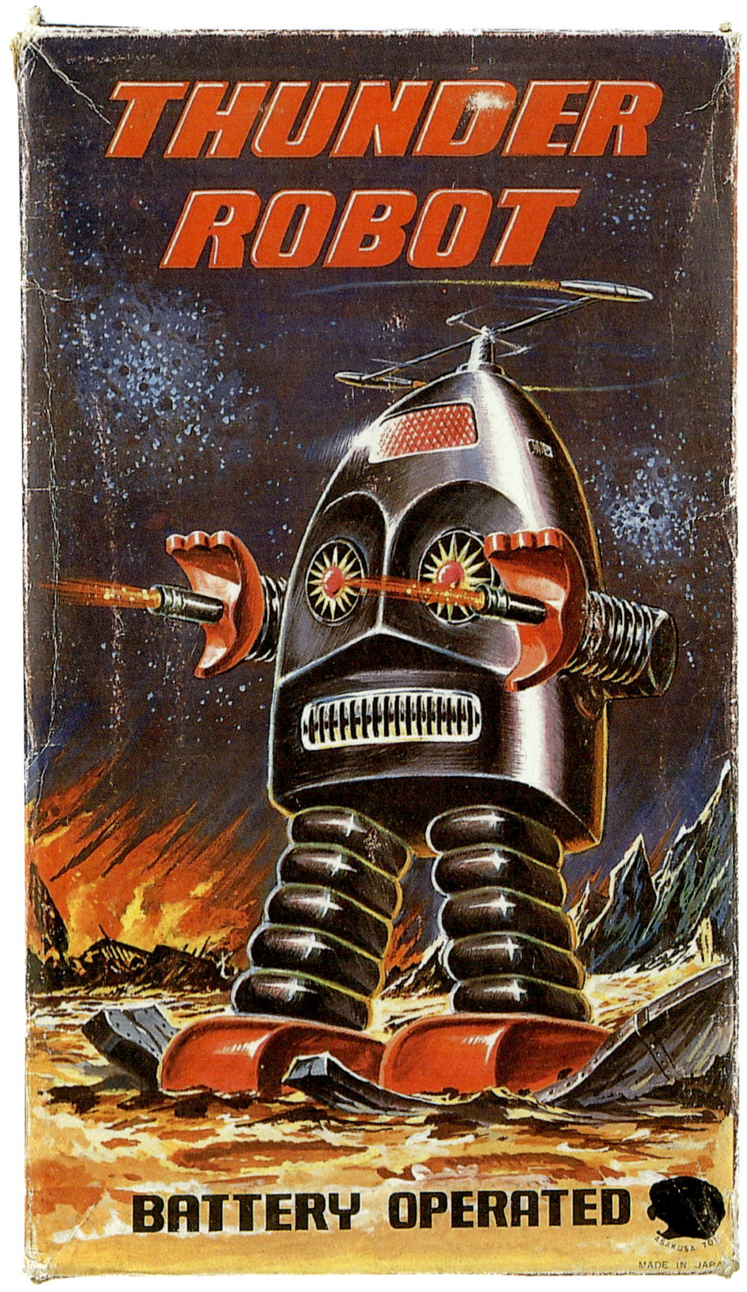

1950s *Thunder Robot*, Asakusa Toy, 90 x 160 x 285 mm

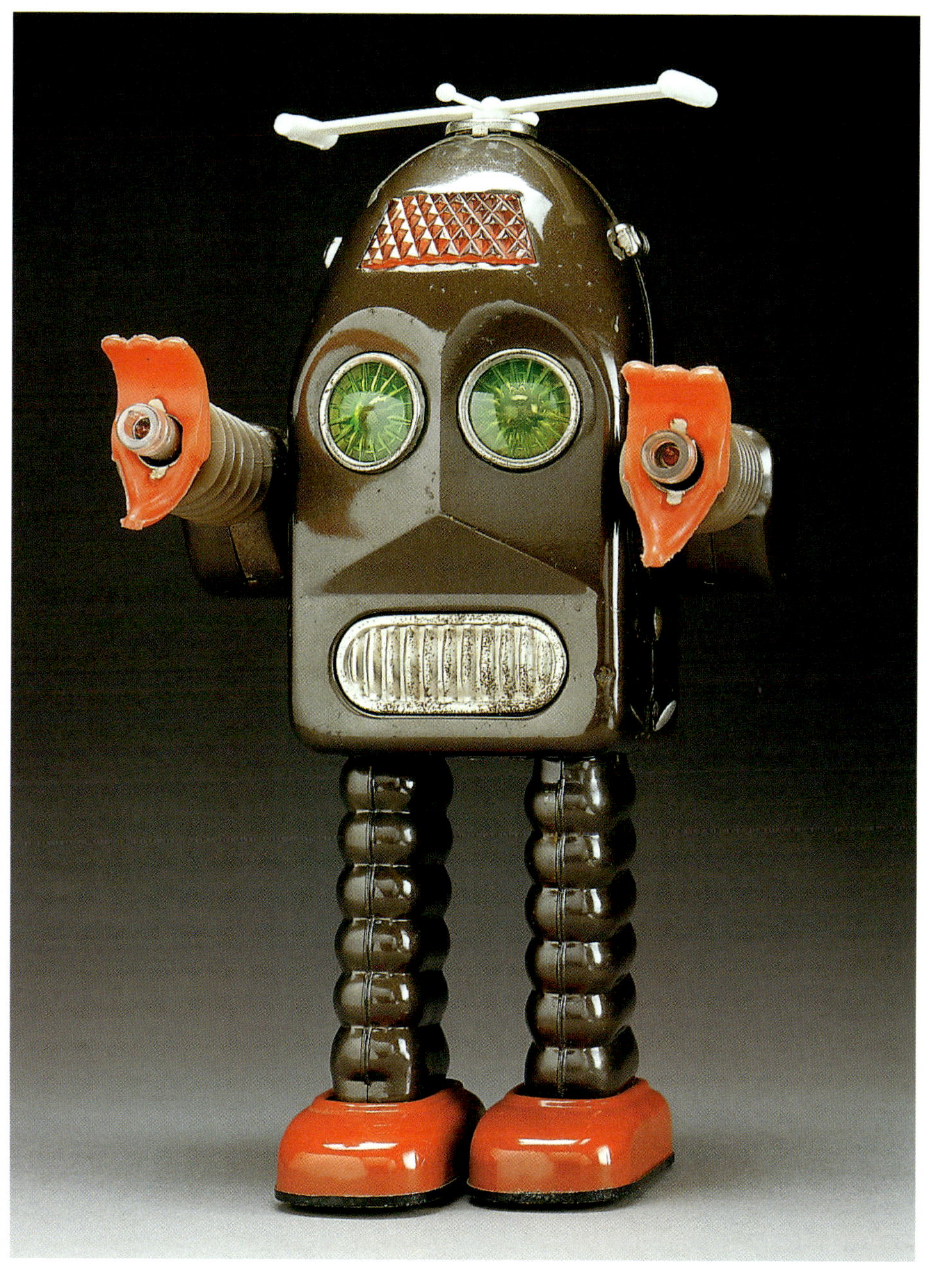

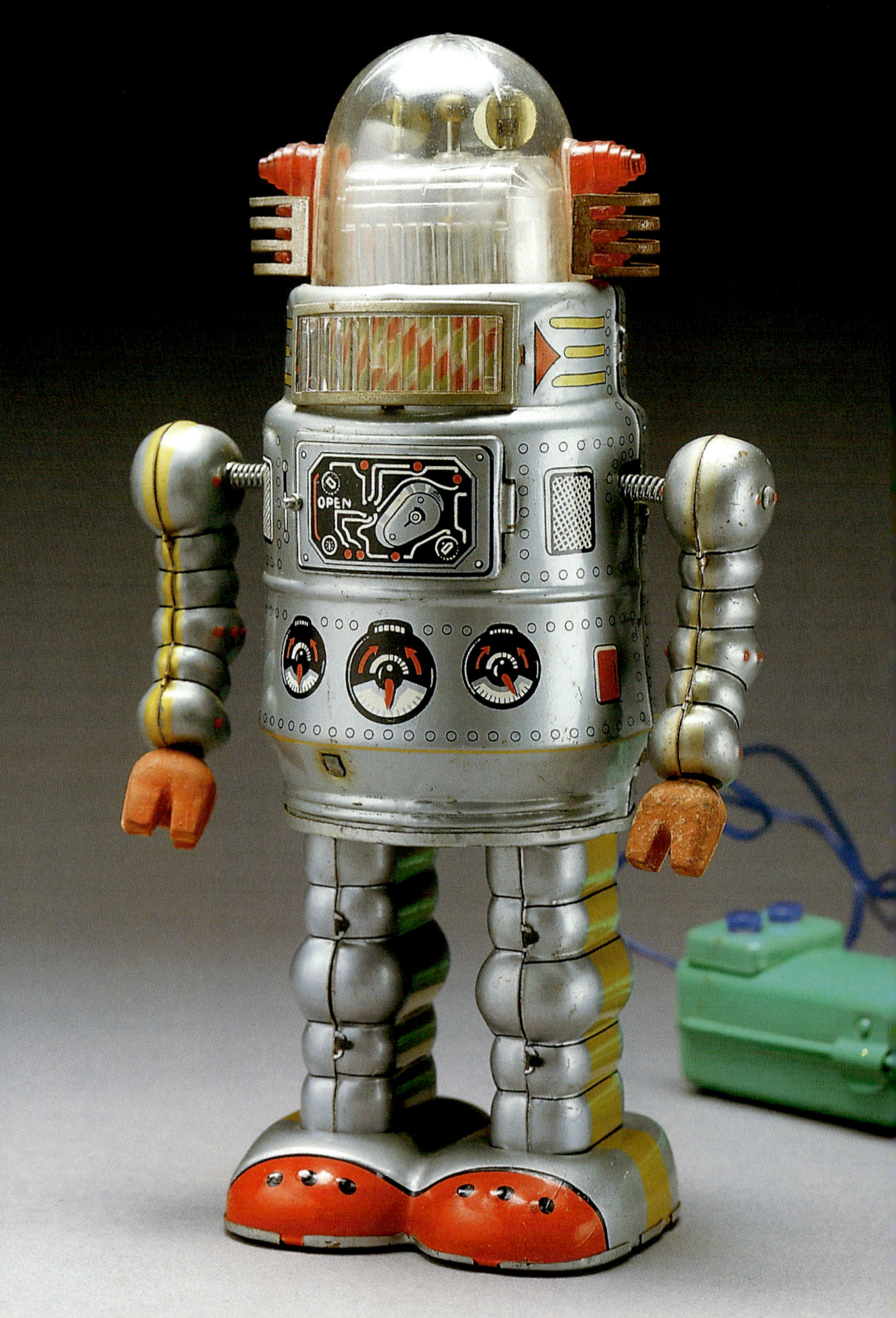

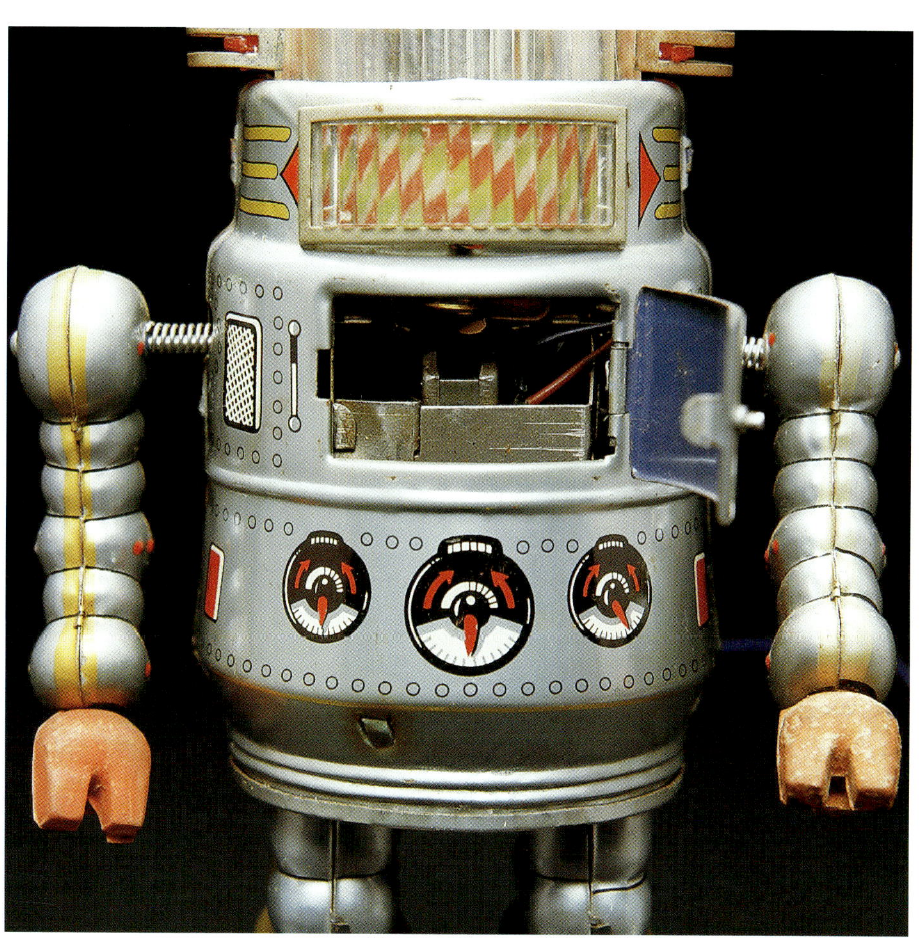

1950s, *Robot*, unknown, 75 x 125 x 230 mm

1950s, *Robot*, Masudaya, 150 x 220 x 380 mm ▶

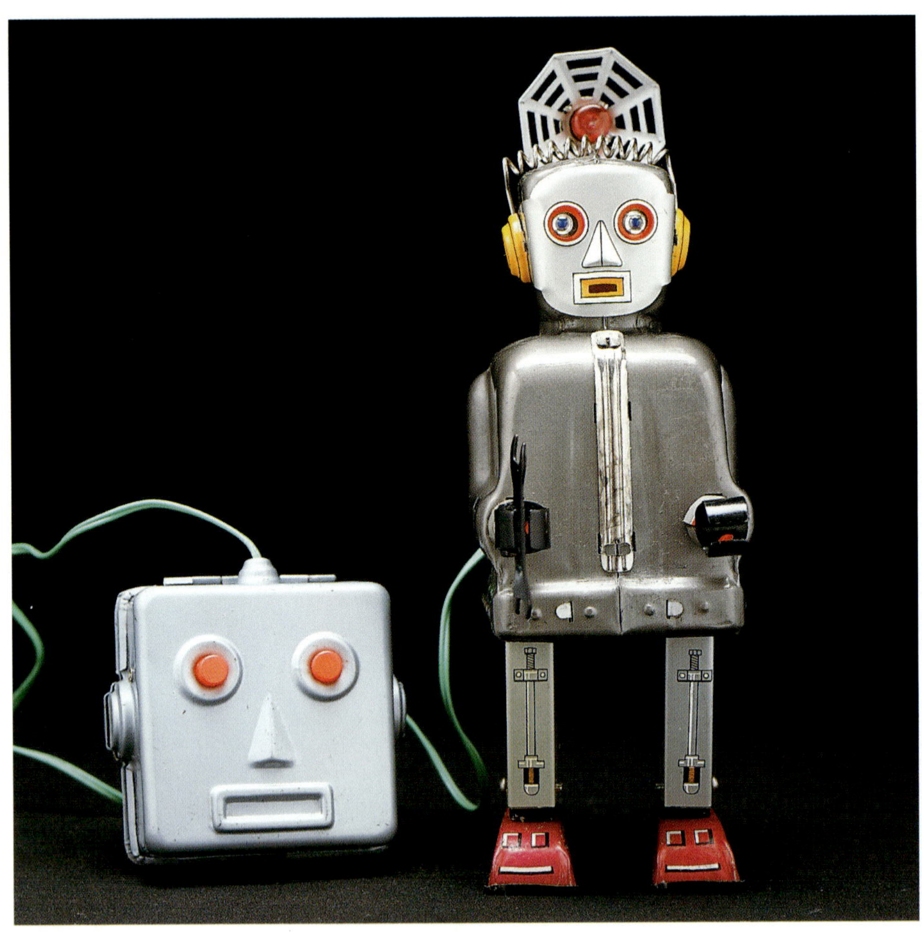

▲ 1950s, *Radar Robot*, Nomura, 100 x 70 x 225 mm

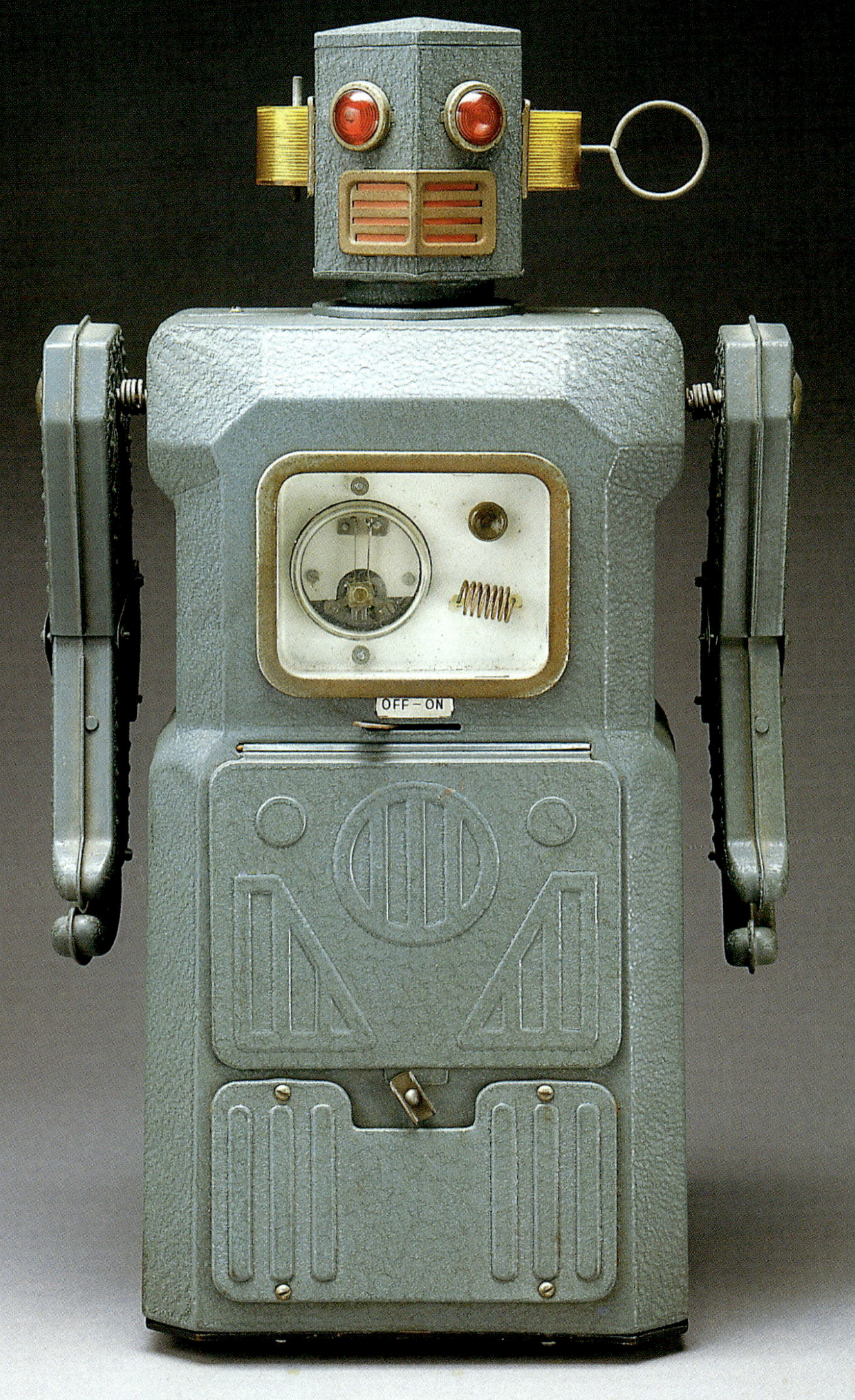
OFF — ON

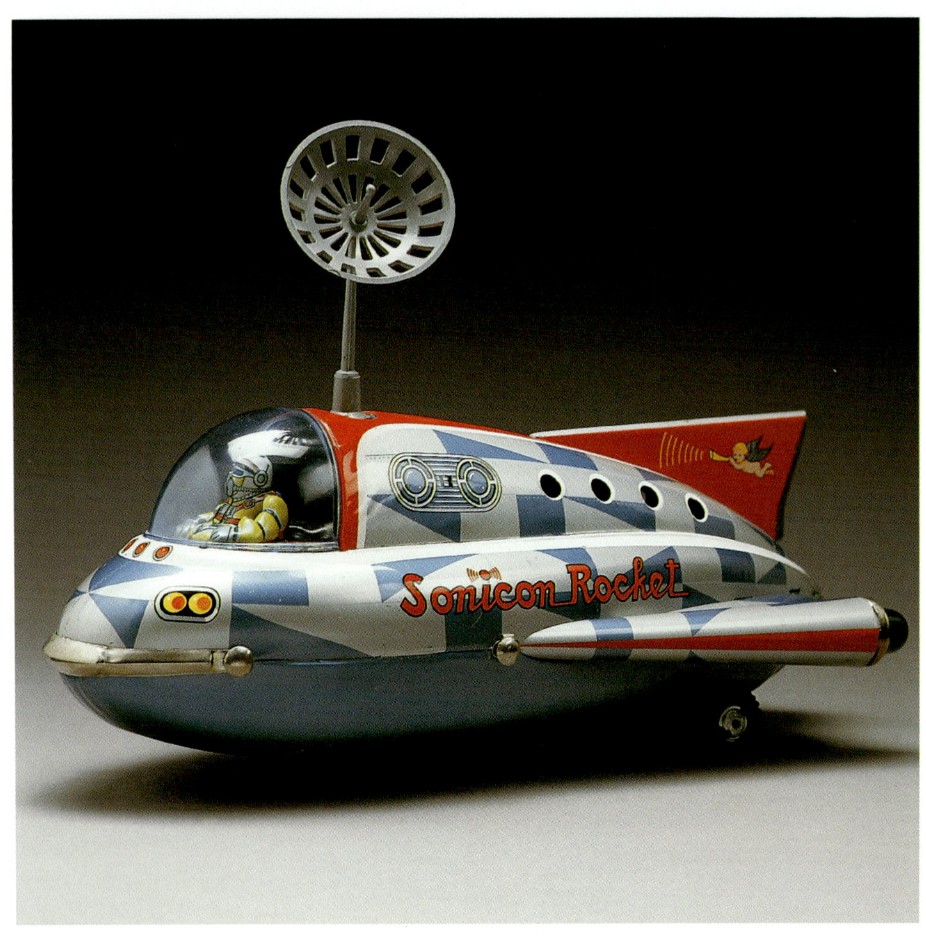

1950s, *Sonicon Rocket*, Masudaya, 340 x 170 x 230 mm

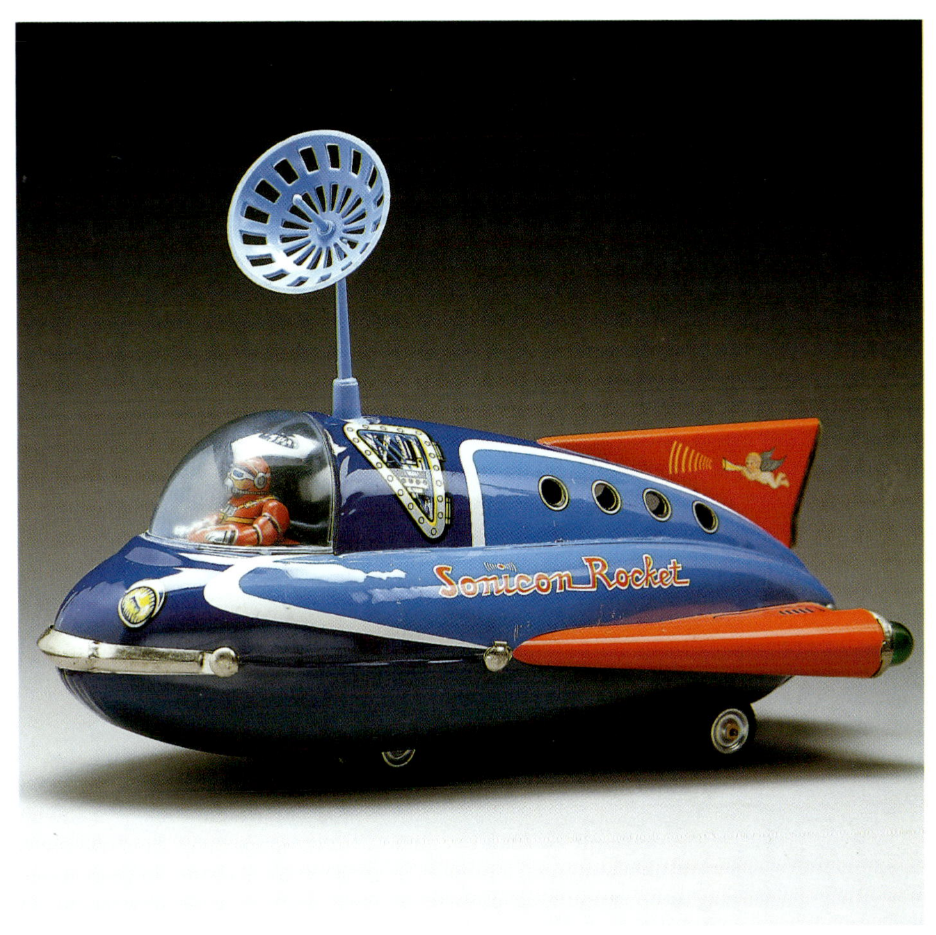

1950s, *Sonicon Rocket*, Masudaya, 340 x 170 x 230 mm

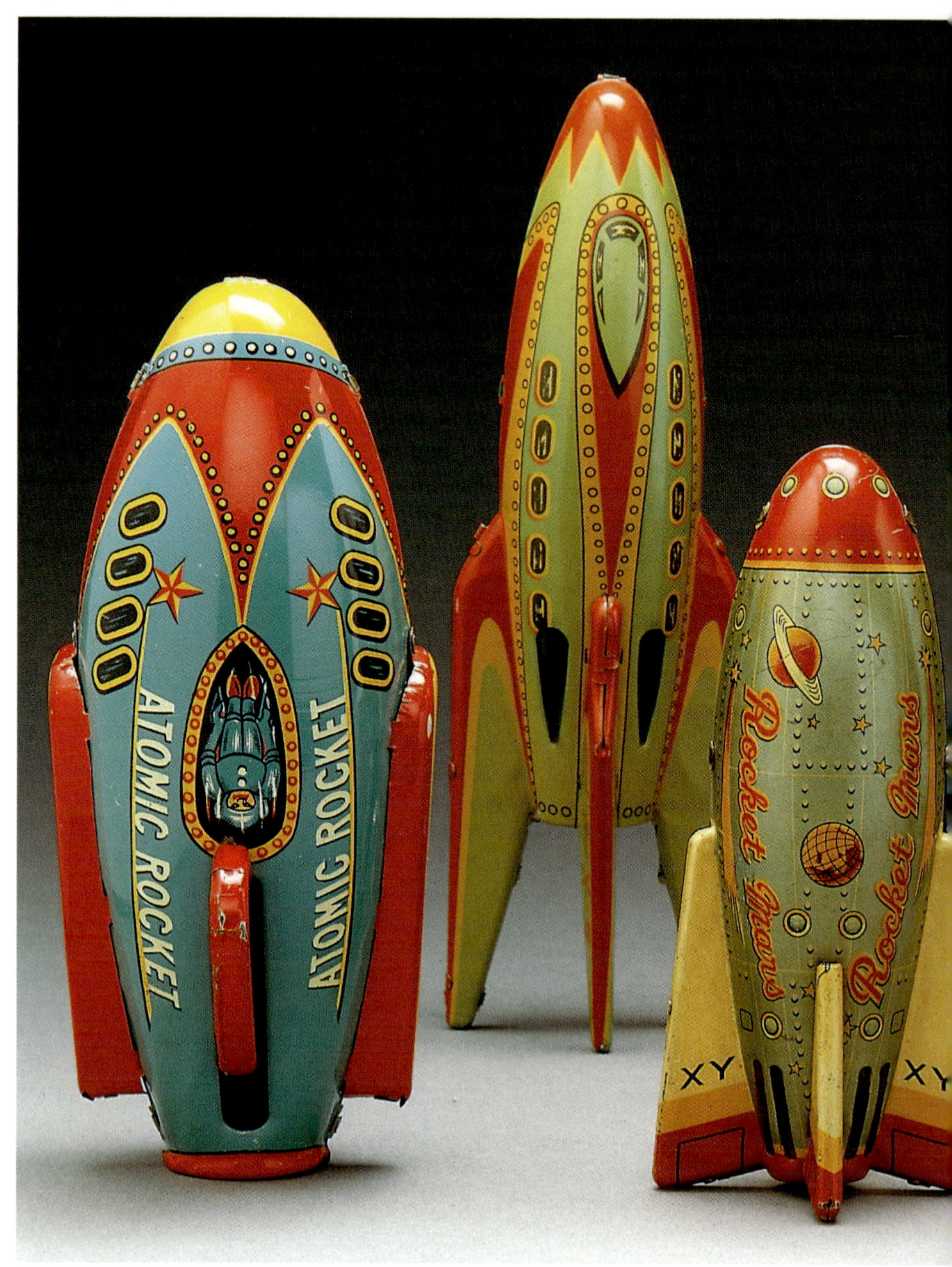

1950s, *Various Rockets*, different sizes

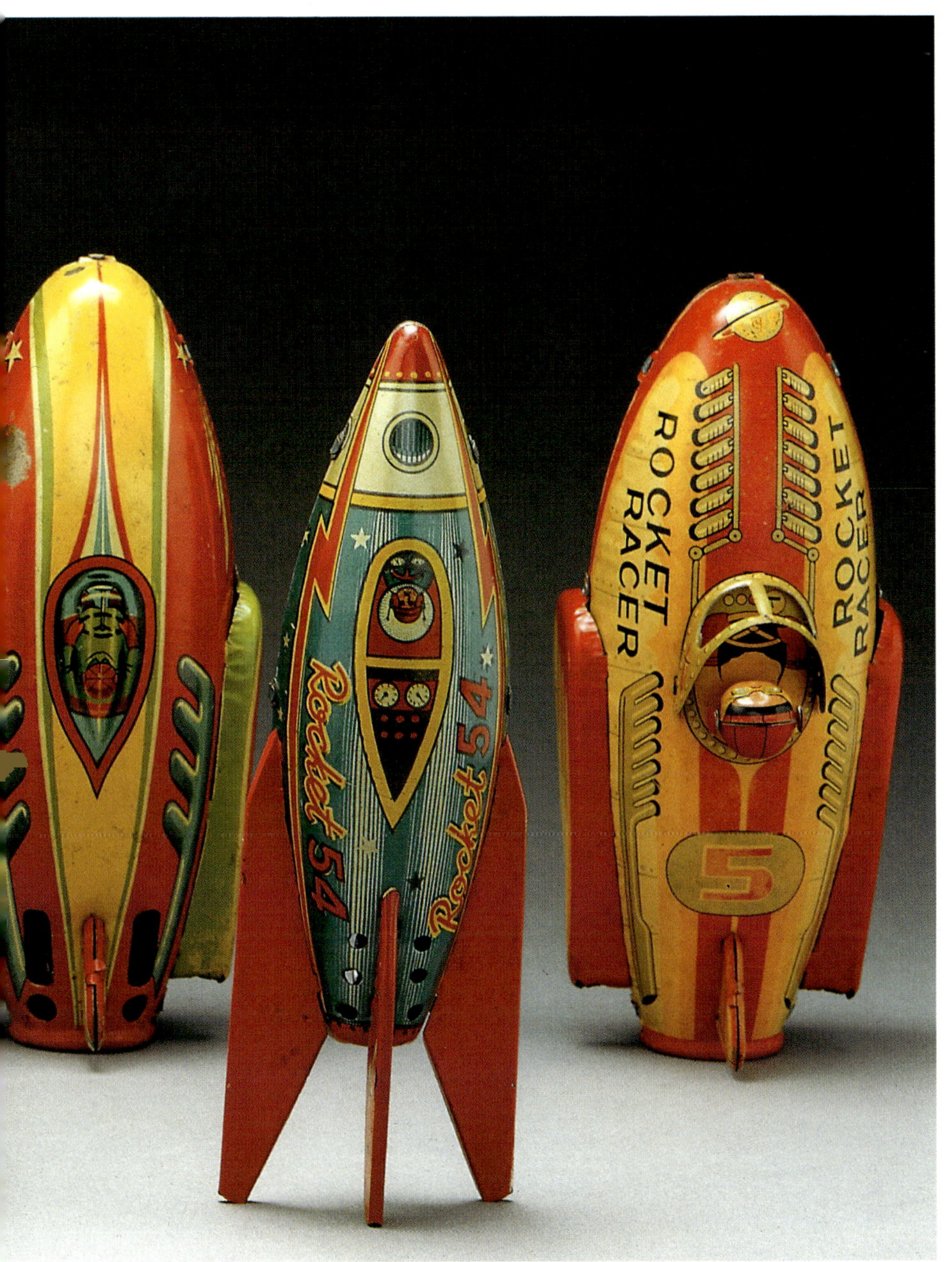
Rocket 54
Rocket 54
ROCKET RACER
ROCKET RACER
5

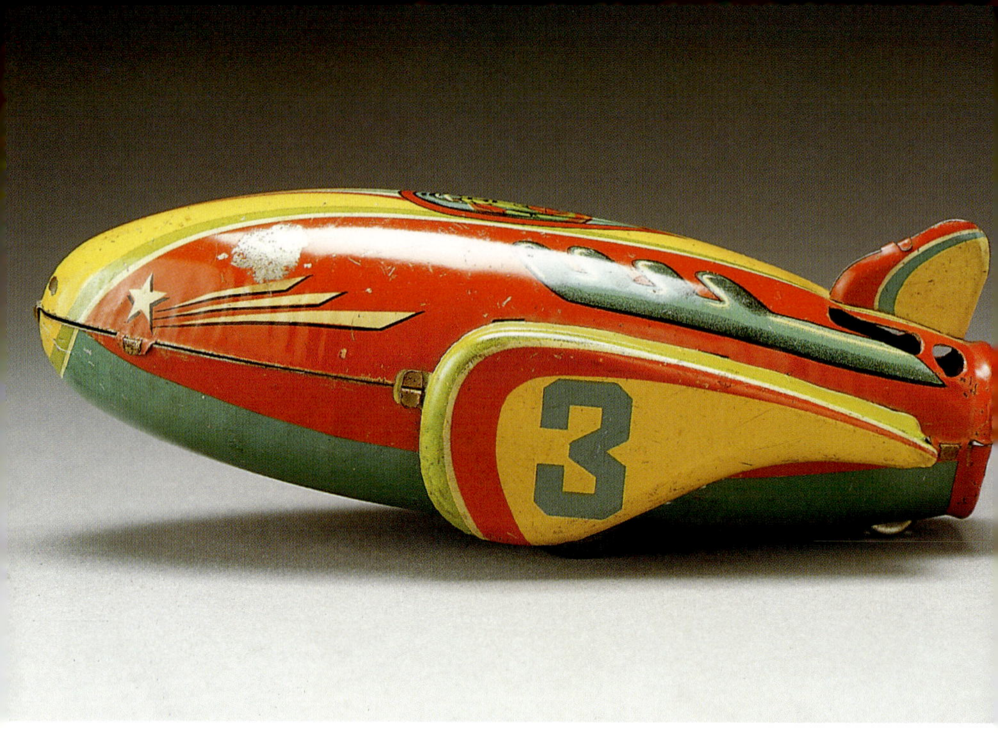

1950s, *Rocket No. 3*, Masudaya, 165 x 70 x 60 mm

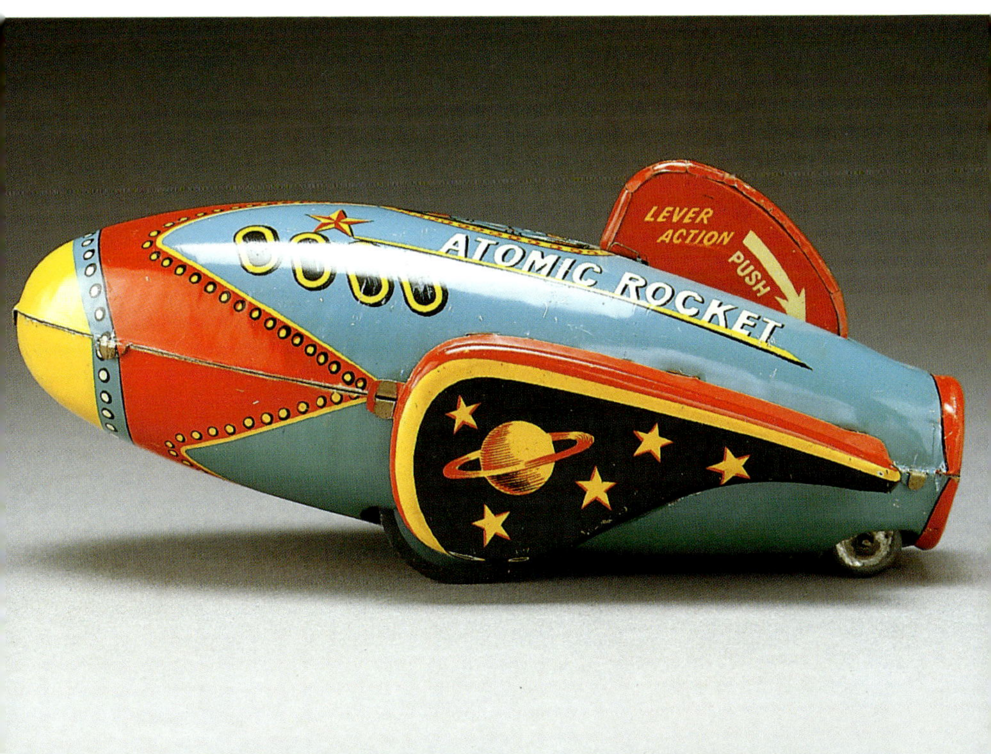

1950s, *Atomic Rocket*, Masudaya, 170 x 70 x 72 mm

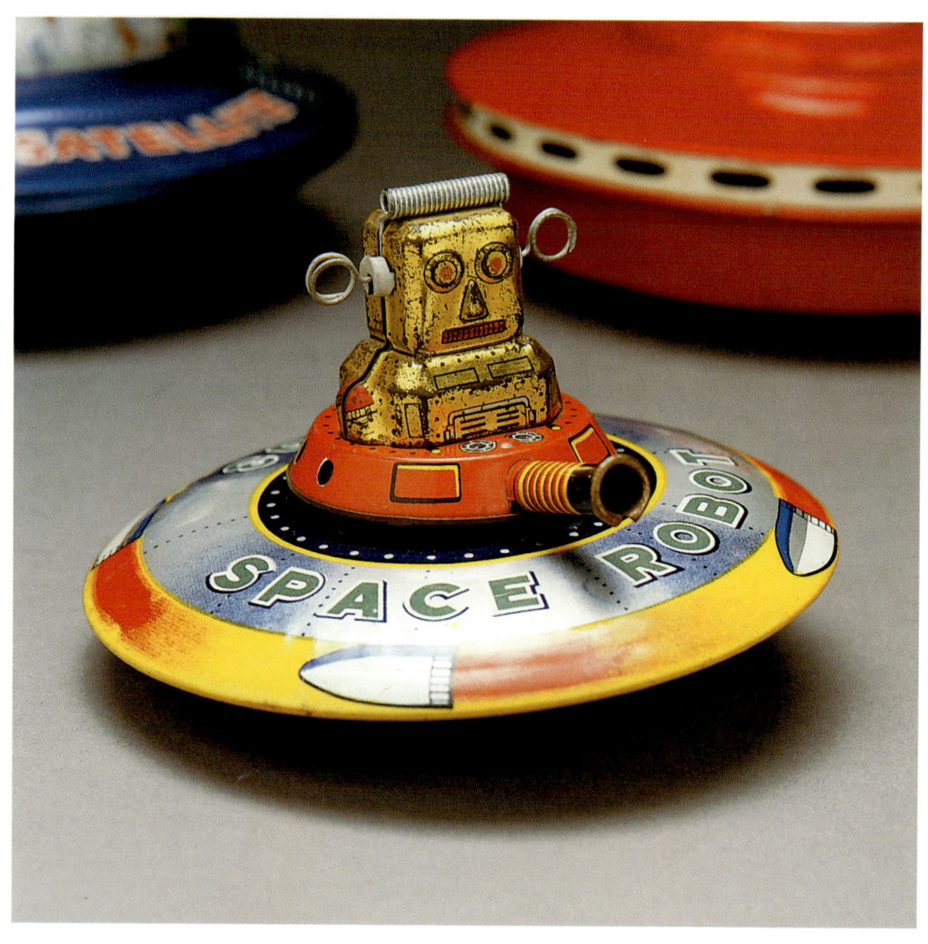

1950s, *Space Robot*, Asahi Toy, 127 x 127 x 80 mm

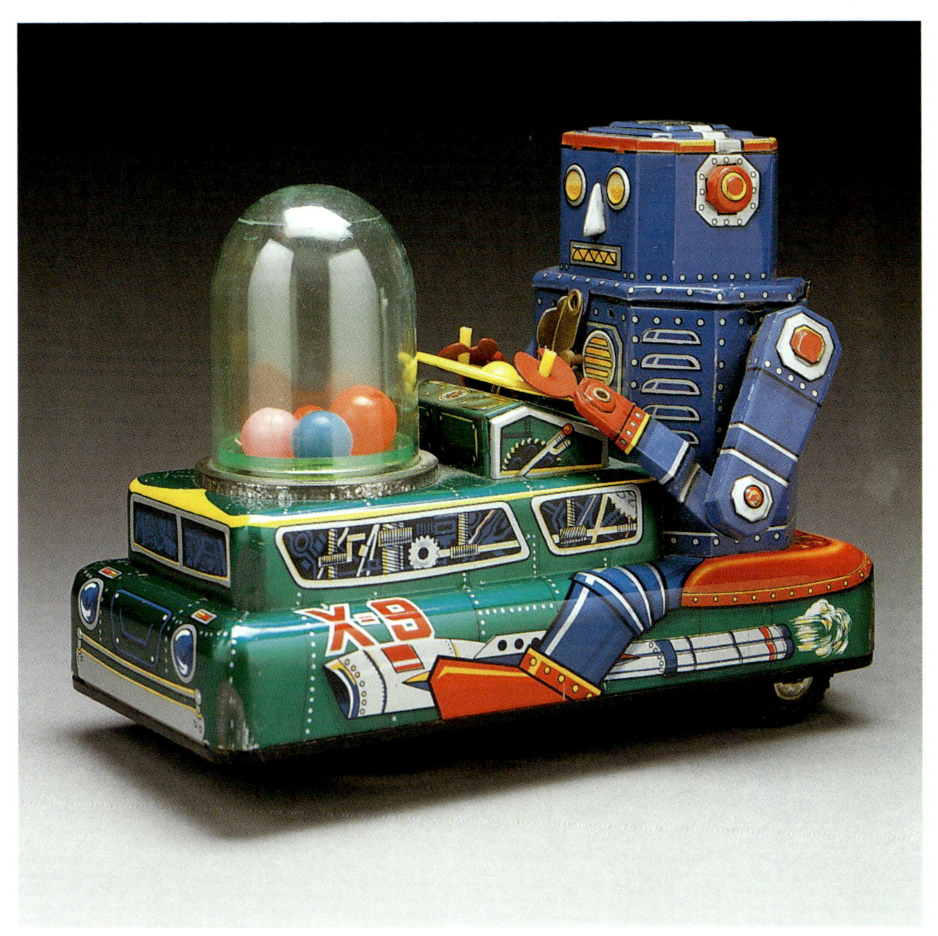

1950s, *X-9 Robot*, Masudaya, 190 x 115 x 155 mm

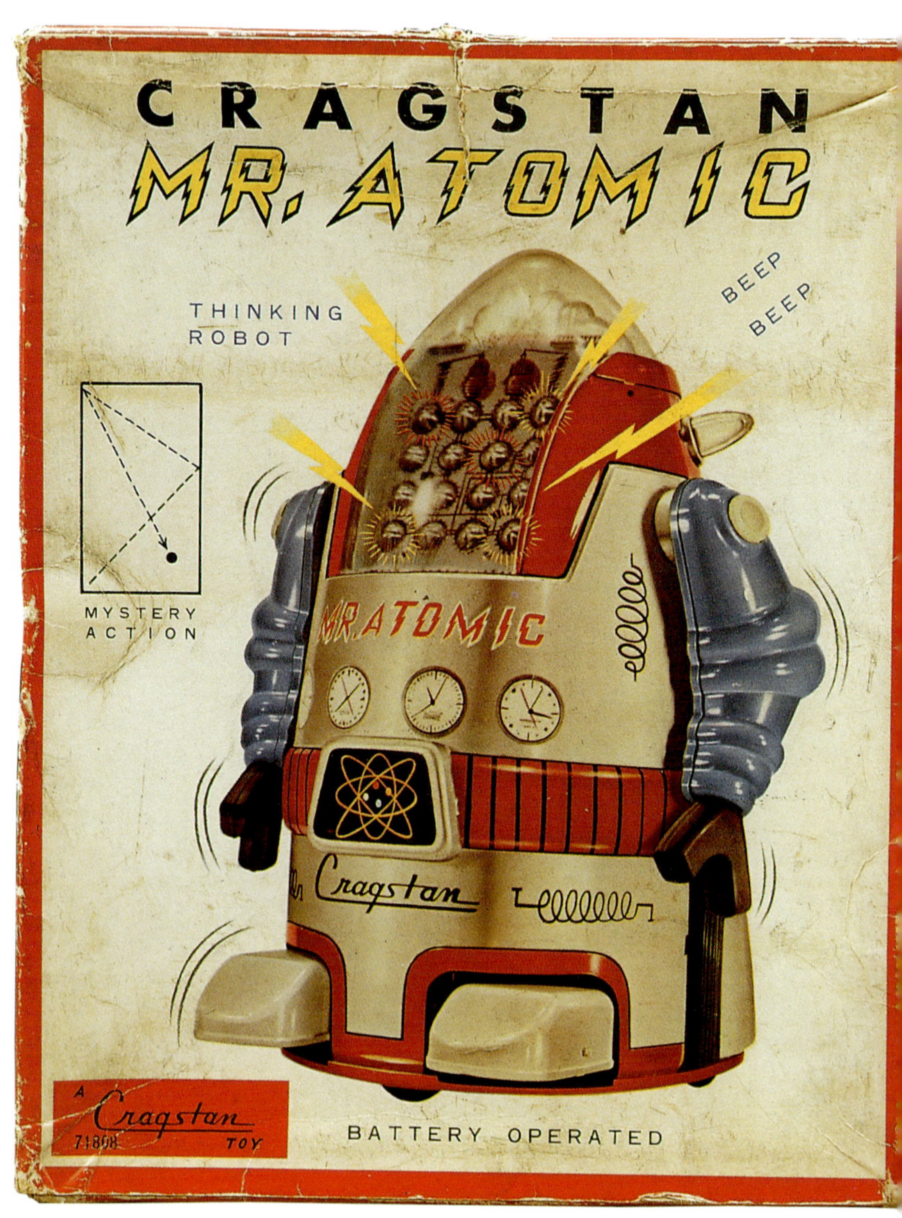

1950s, *Mr. Atomic*, Yonezawa, 155 x 175 x 225 mm

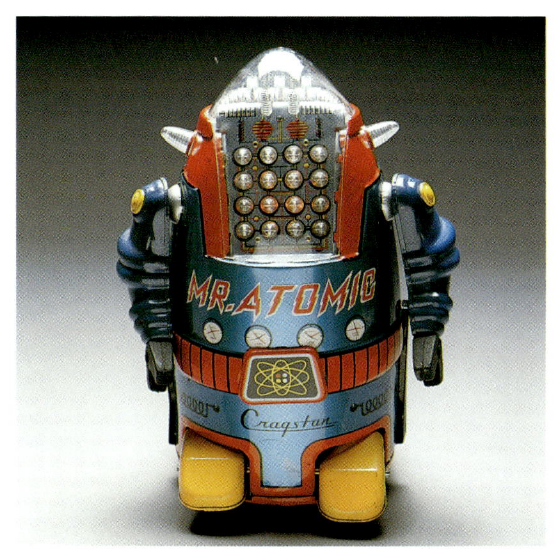

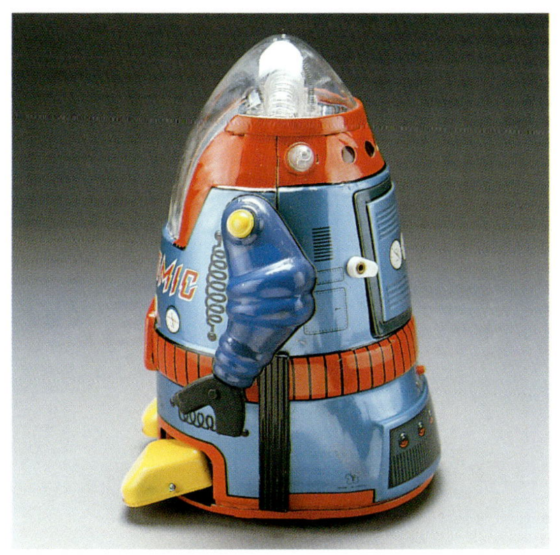

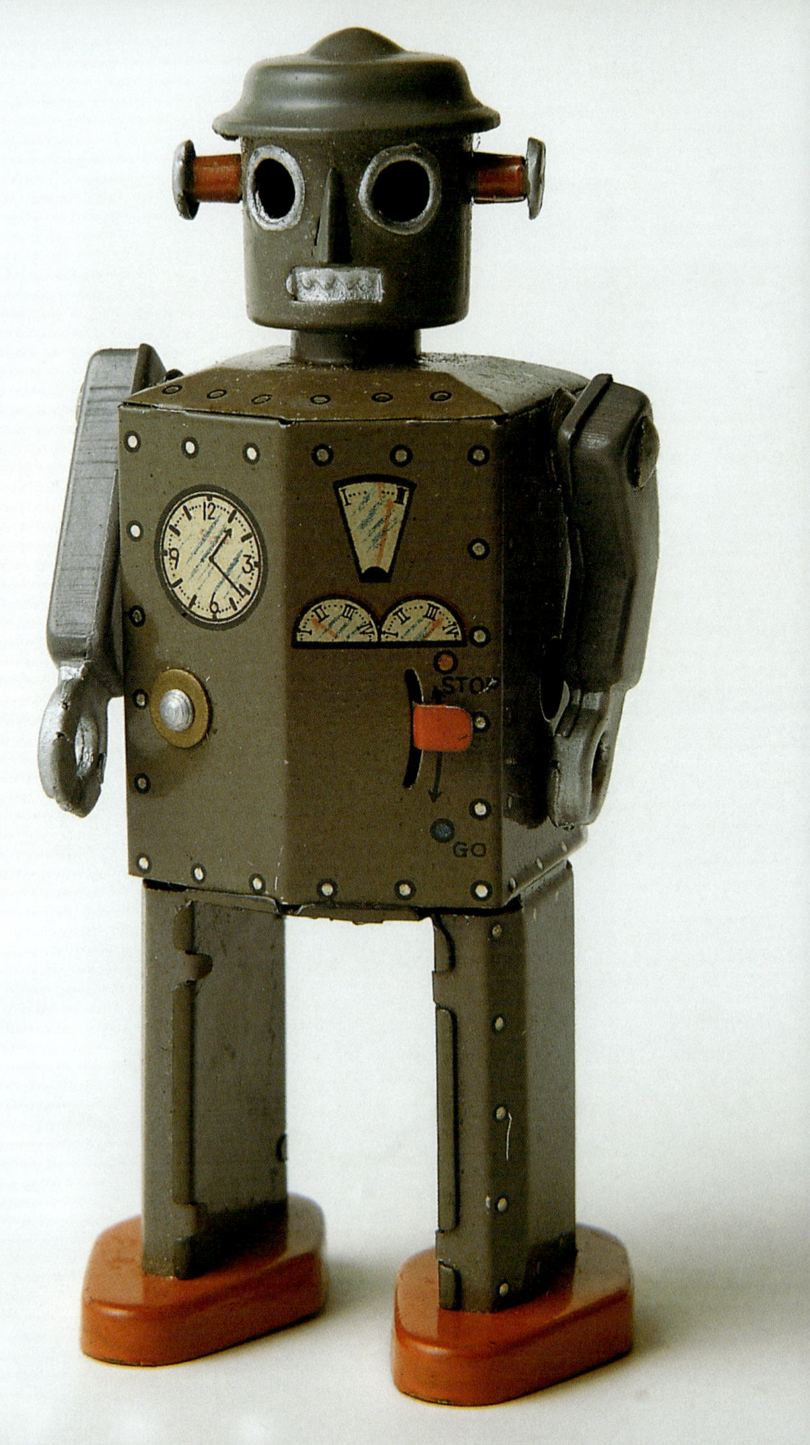

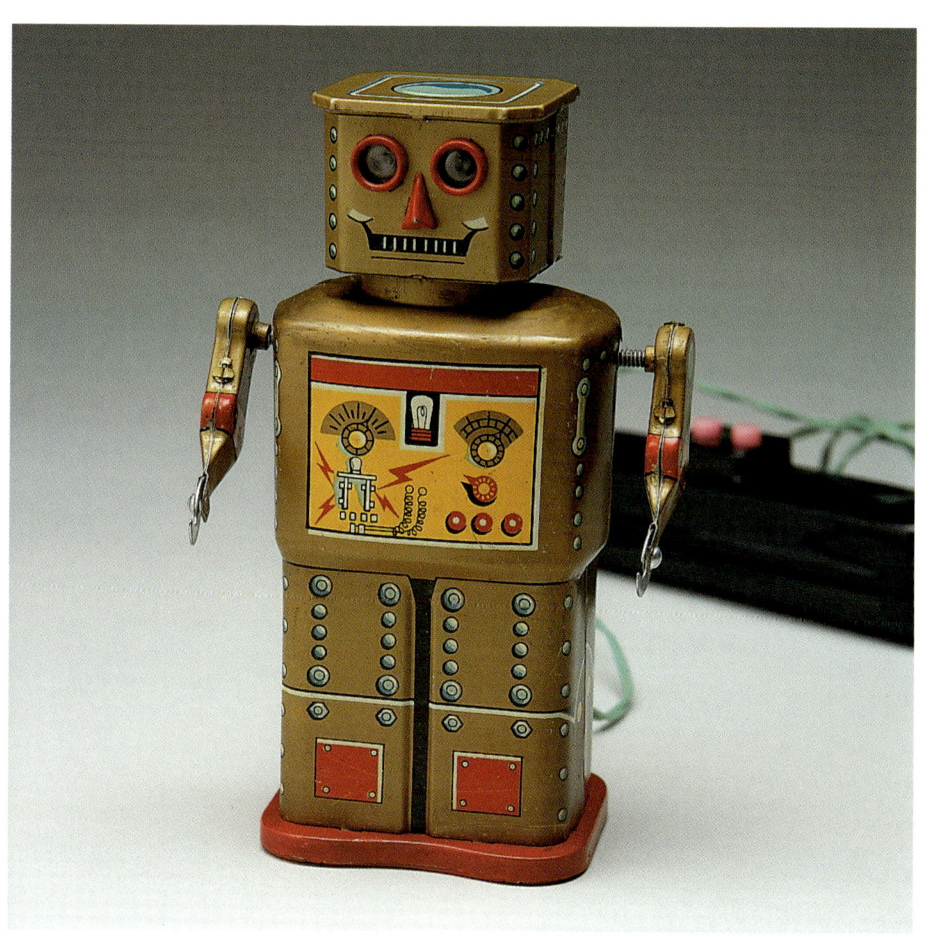

▲ 1950s, *Robot*, Line Mar, 55 x 110 x 165 mm

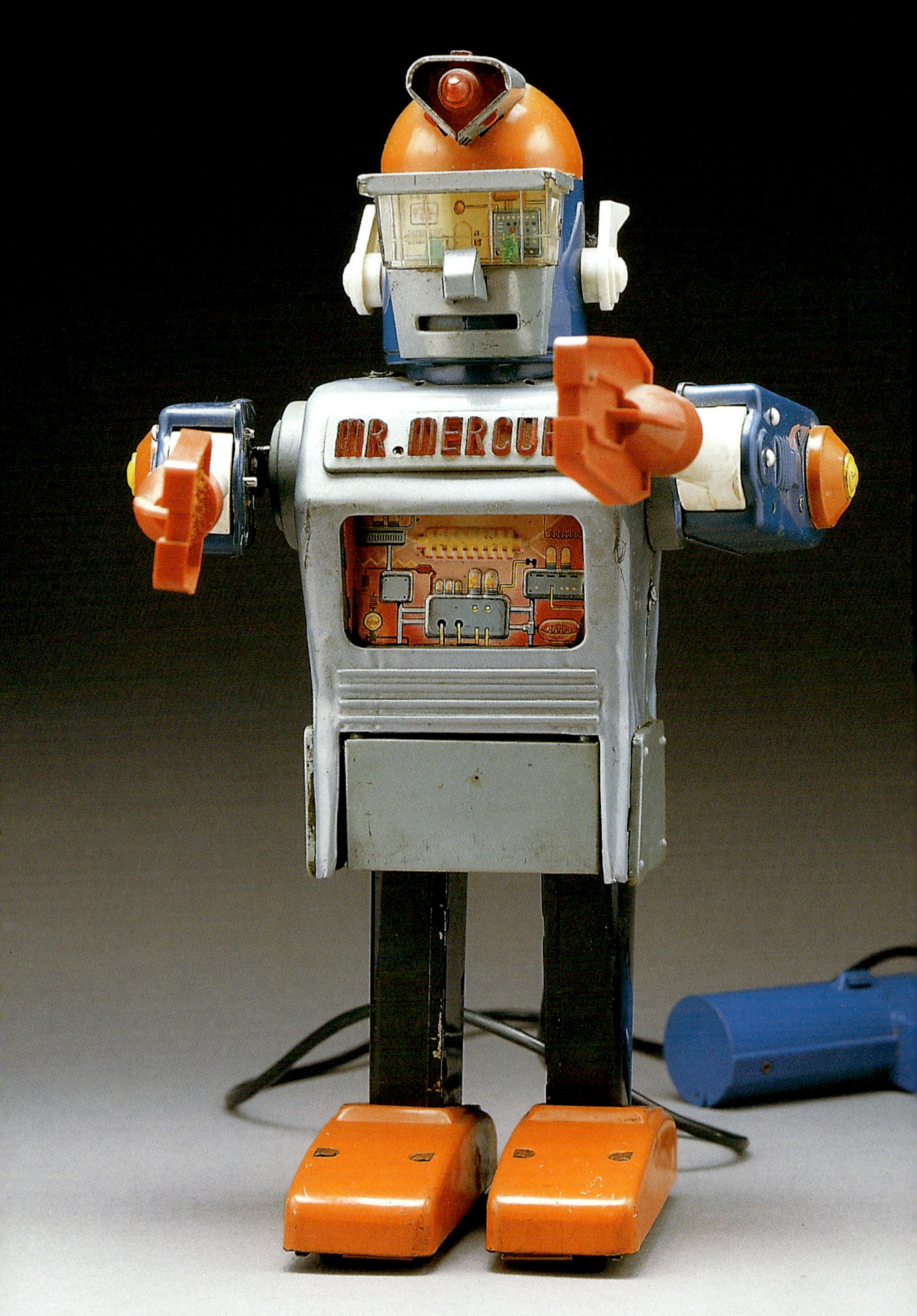

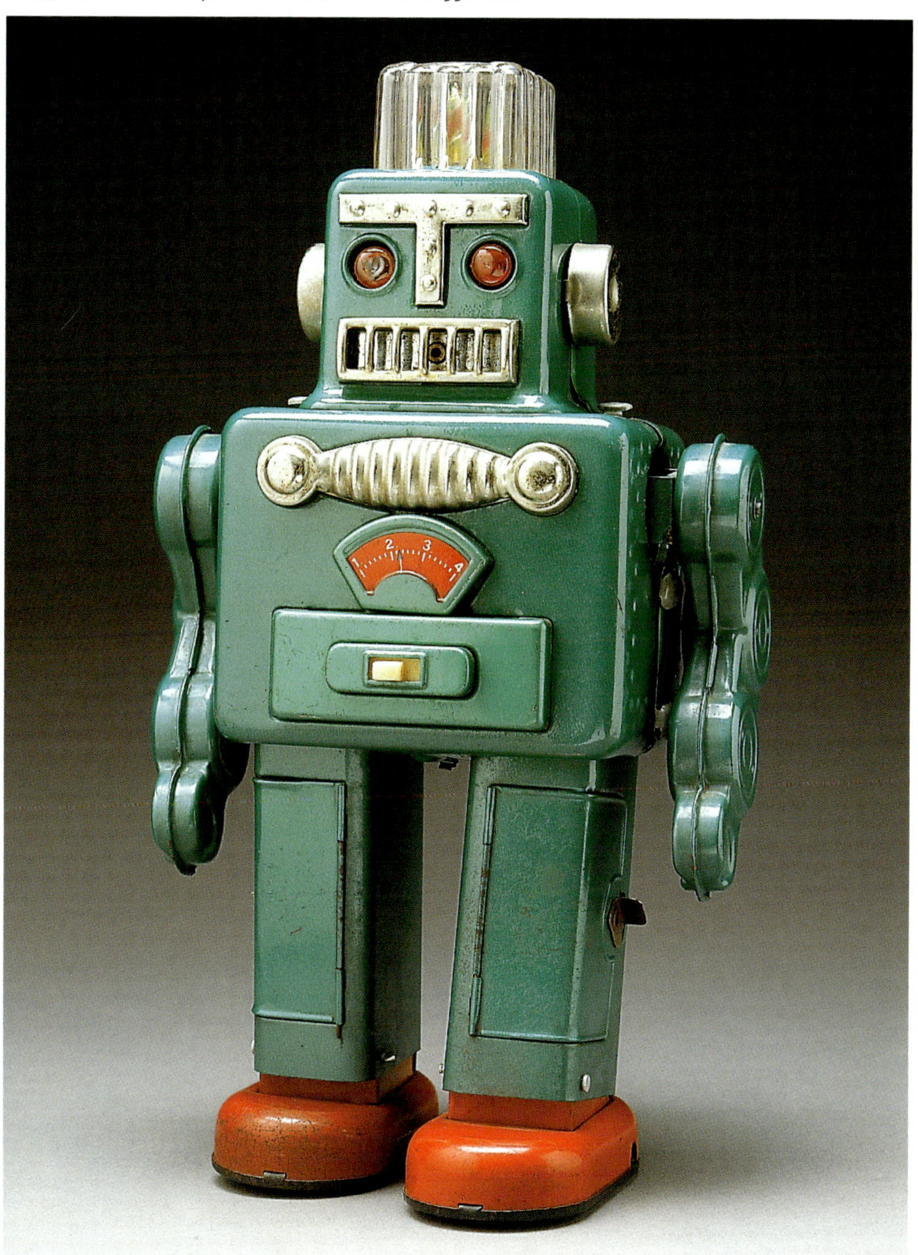

▲1950s, *Smoking Robot*, Yonezawa, 105 x 160 x 300 mm

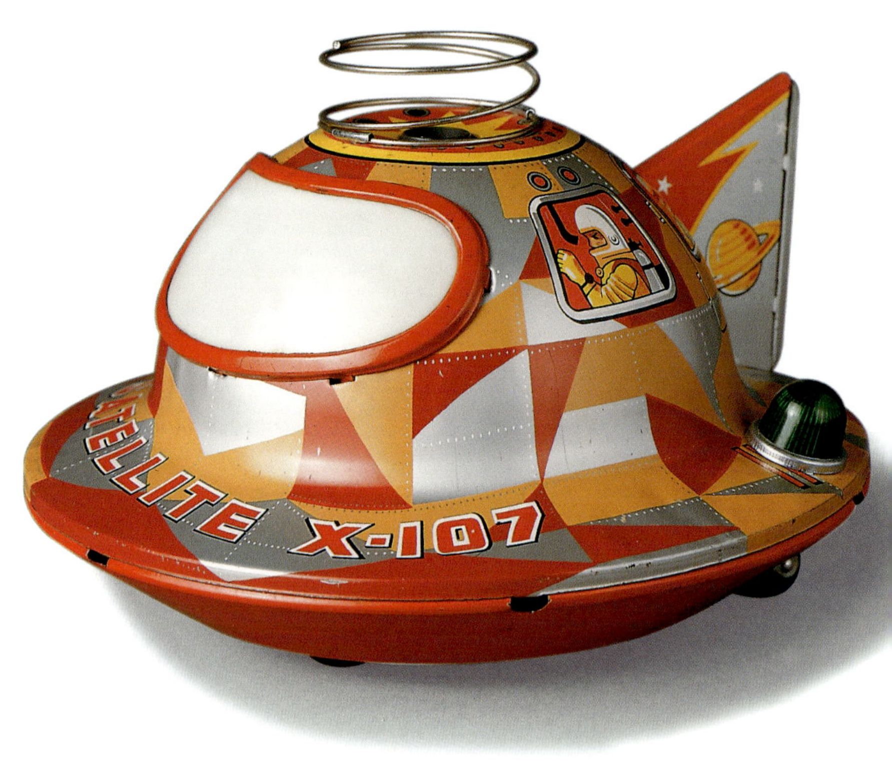

1960s, *Satellite X-107*, Masudaya, 200 x 200 x 128 mm

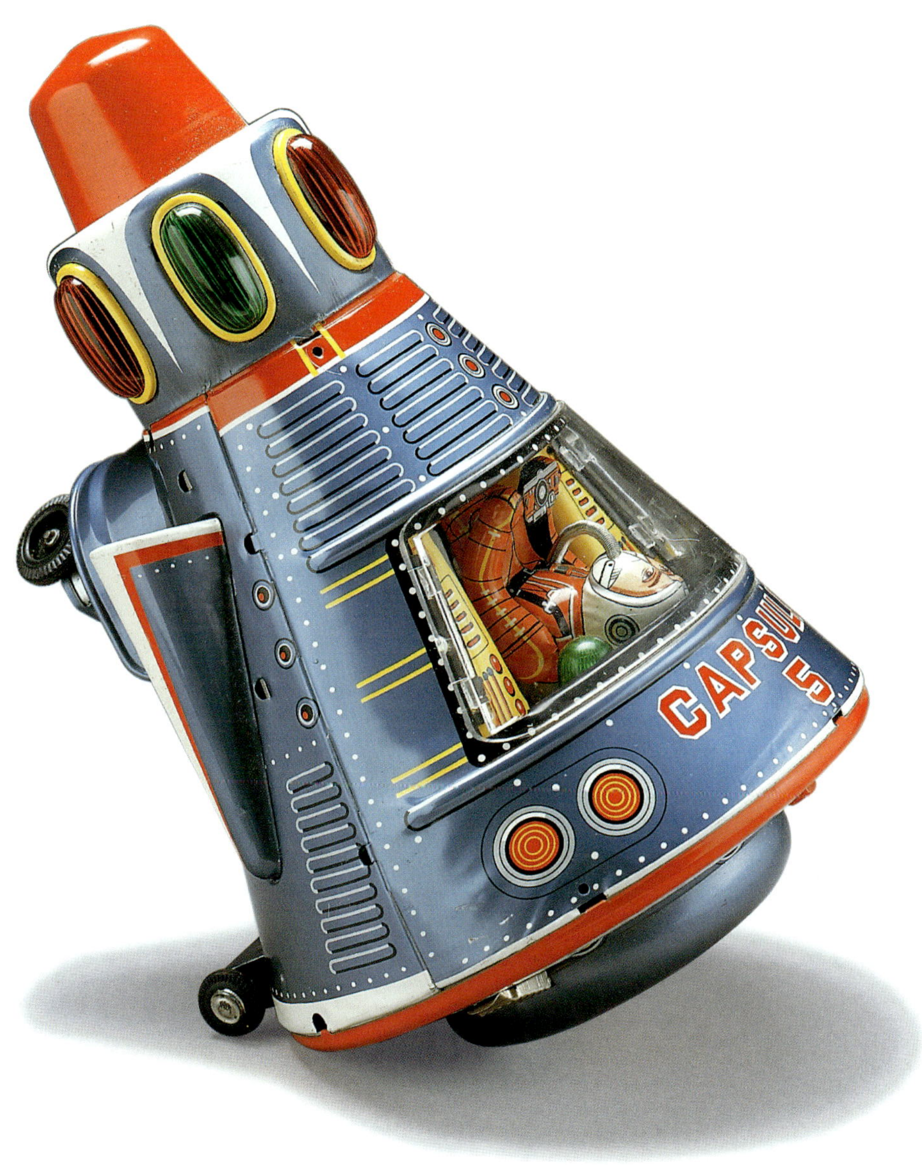

1950s, *Capsule 5*, Masudaya, 265 x 160 x 175 mm

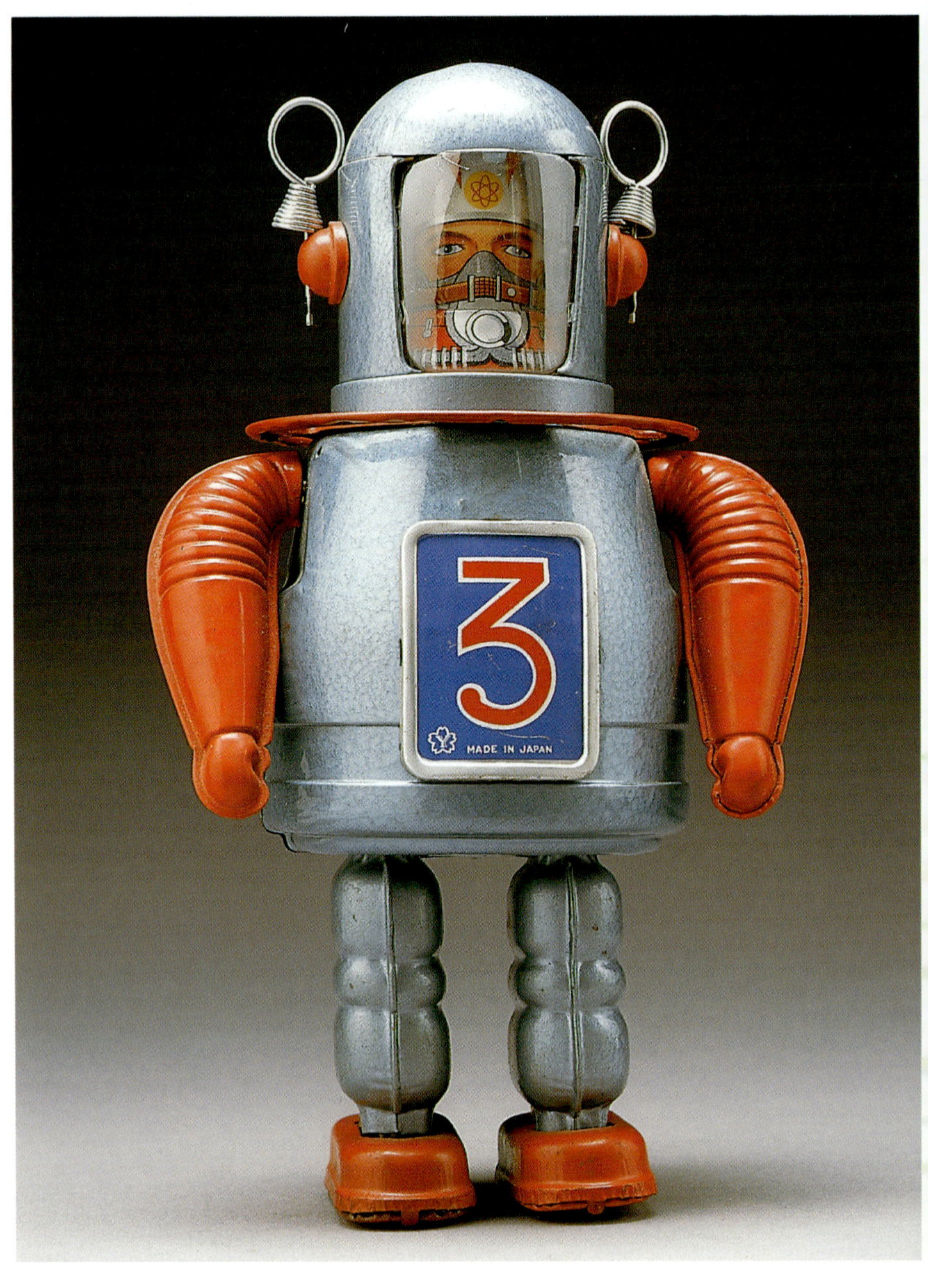

1950s, *Astro Scout*, Yonezawa, 85 x 130 x 230 mm

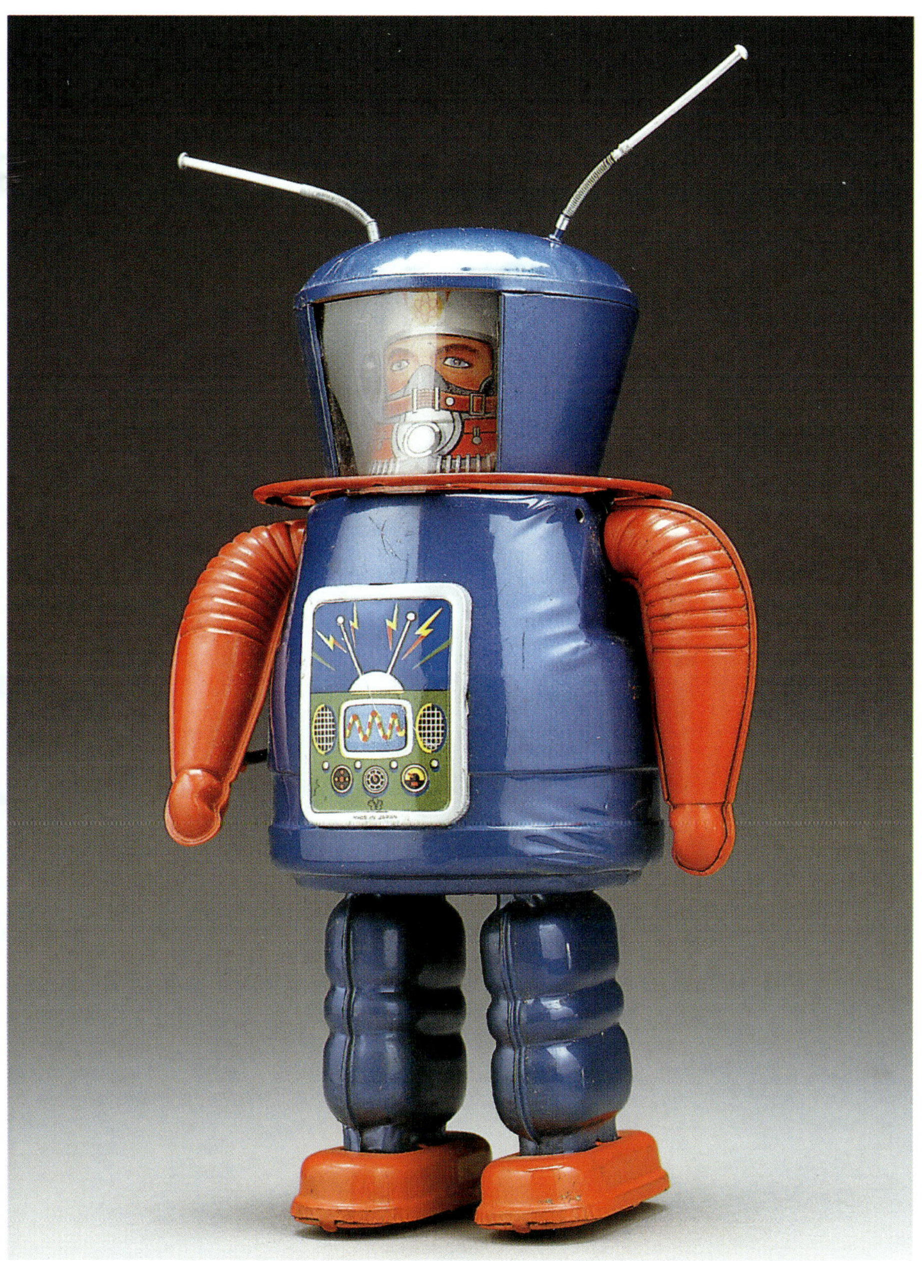

1950s, *X-27 Explorer*, Yonezawa, 85 x 125 x 213 mm

1960s, *Space Frontier*, Yoshino Toy, 445 x 120 x 215 mm

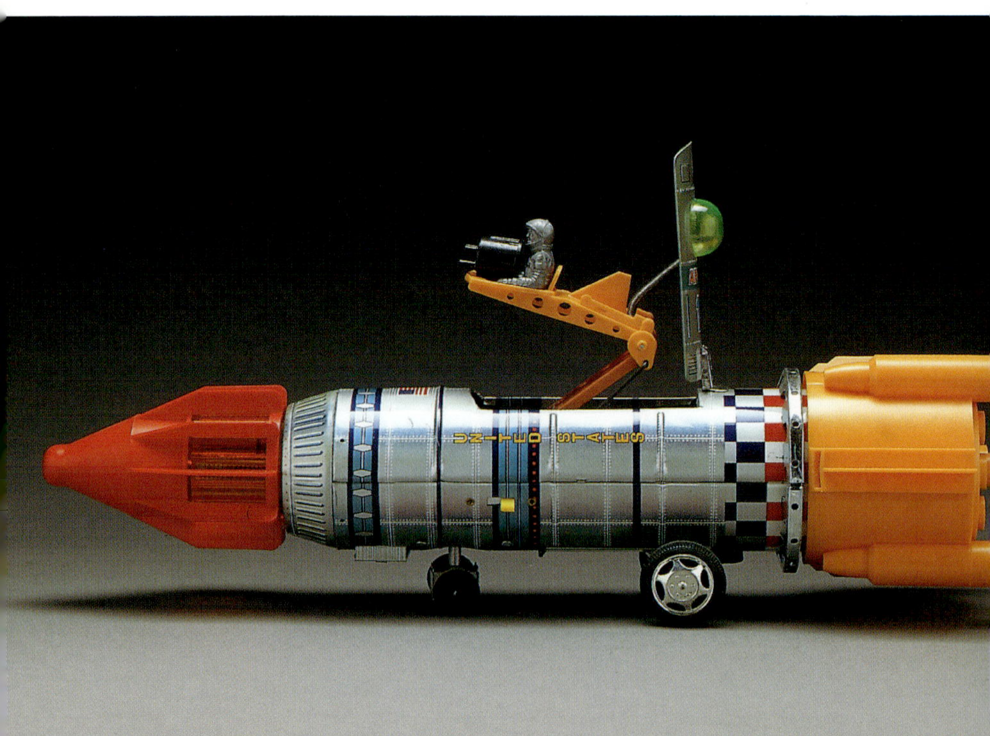

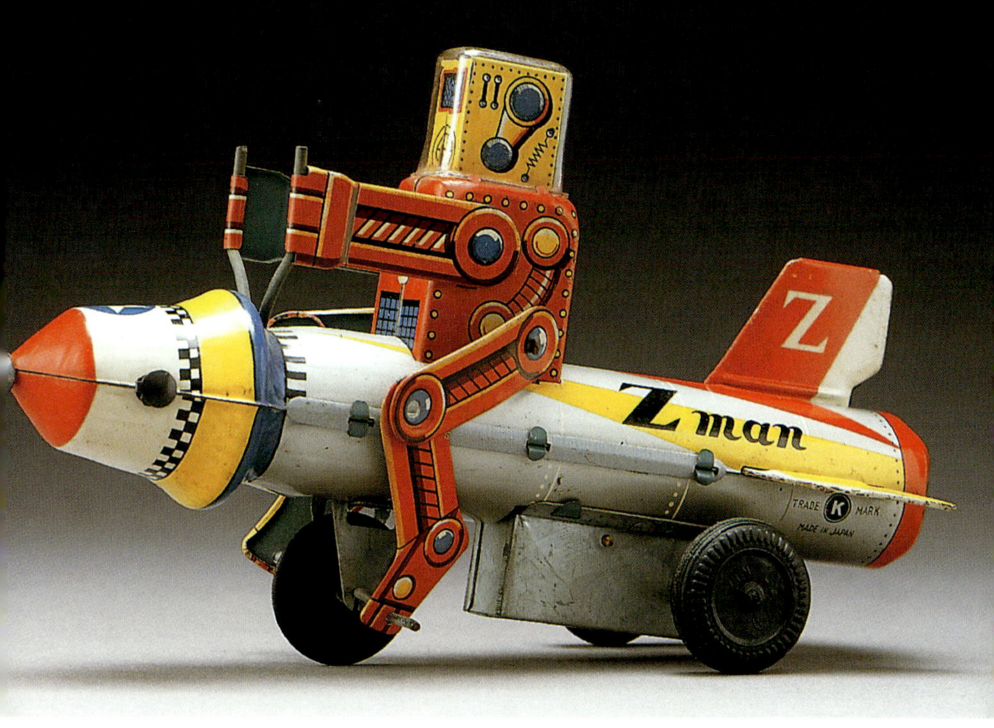

1950s, *Space Patrol*, Ohta, 180 x 63 x 110 mm

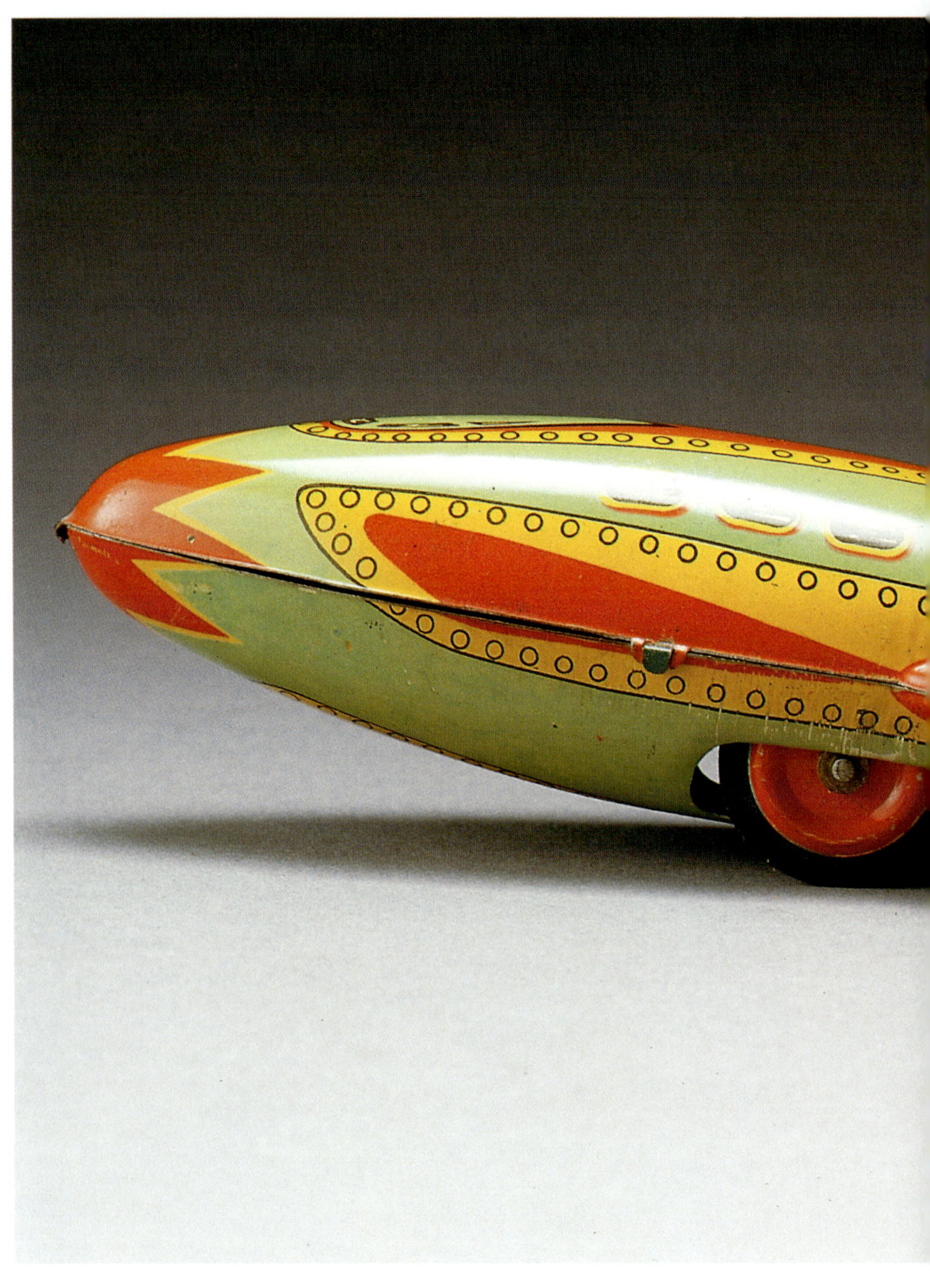

1950s, *Space Ship X-3*, Masudaya, 205 x 62 x 47 mm

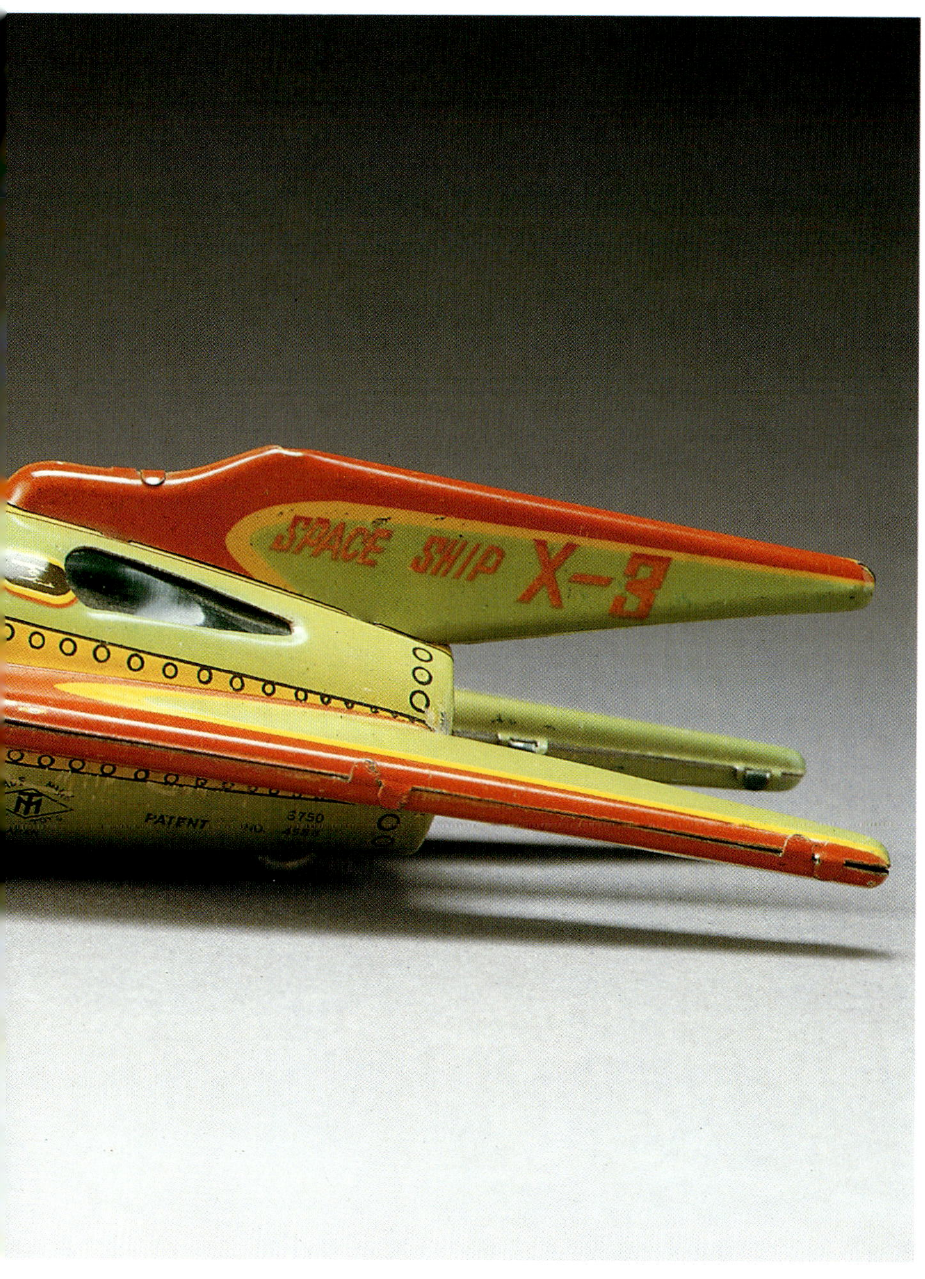
SPACE SHIP X-3
PATENT NO. 3750

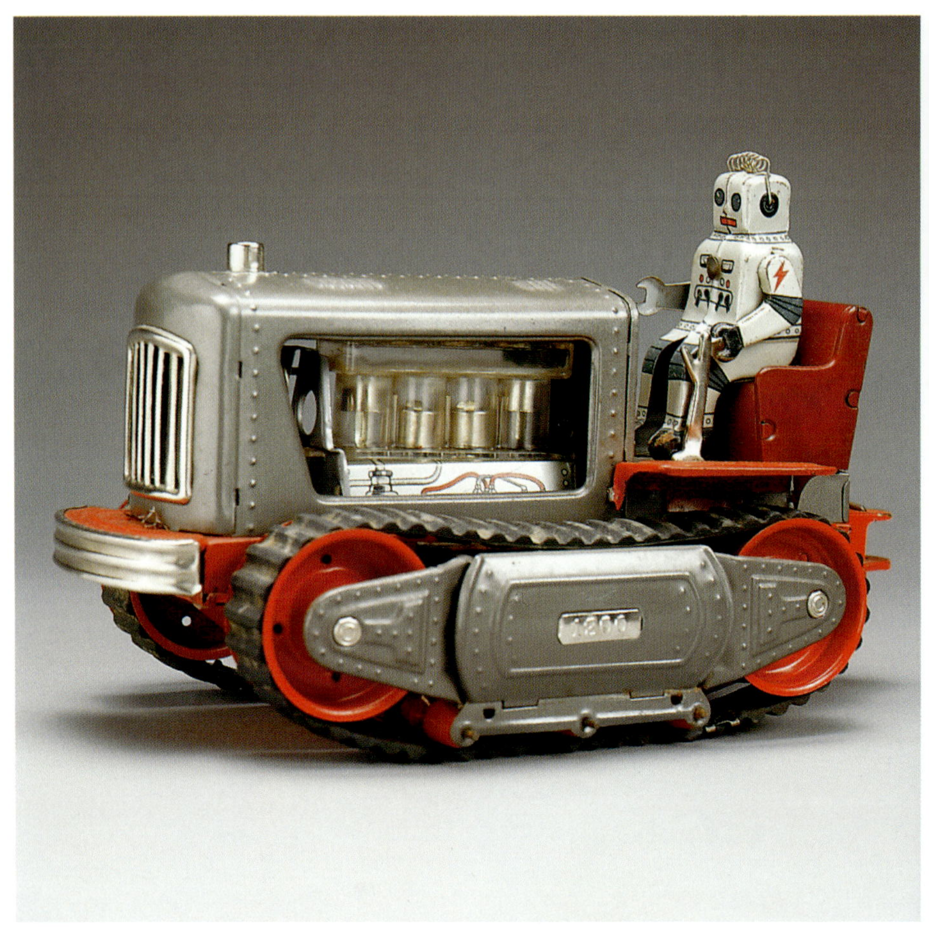

1950s, *Robot Tractor*, Showa, 250 x 125 x 150 mm

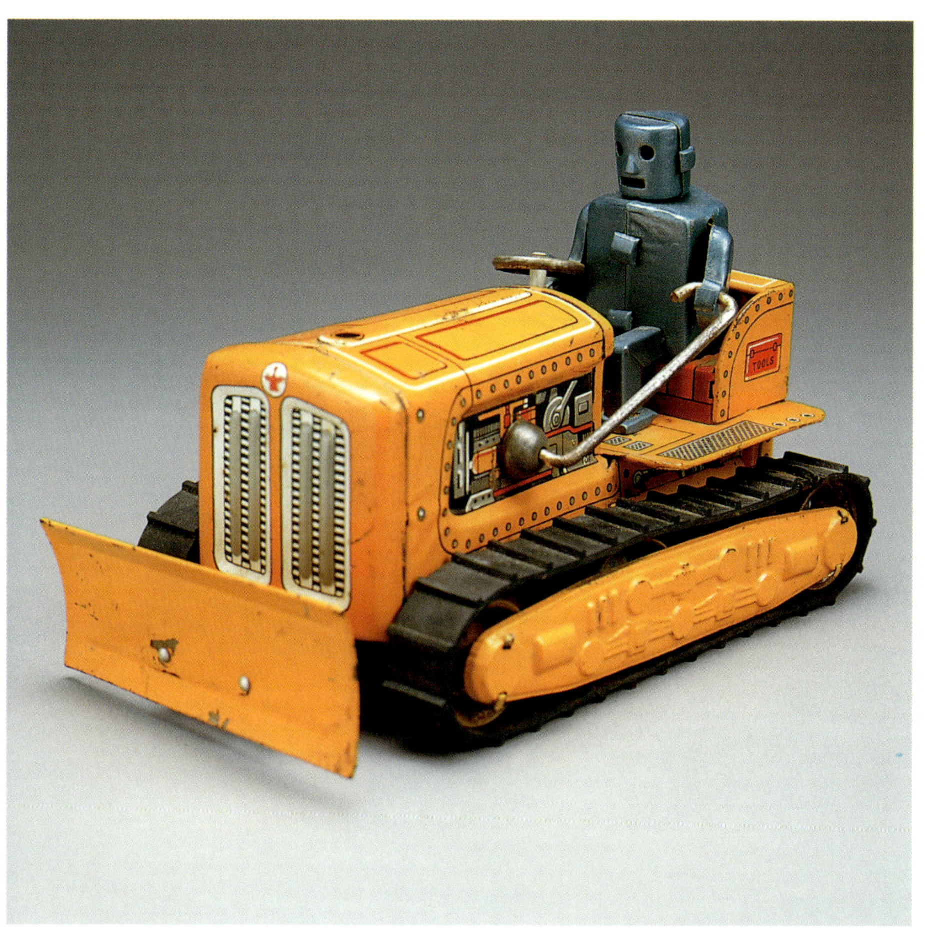

1950s, *Robot Bulldozer*, Yoshiya, 180 x 97 x 120 mm

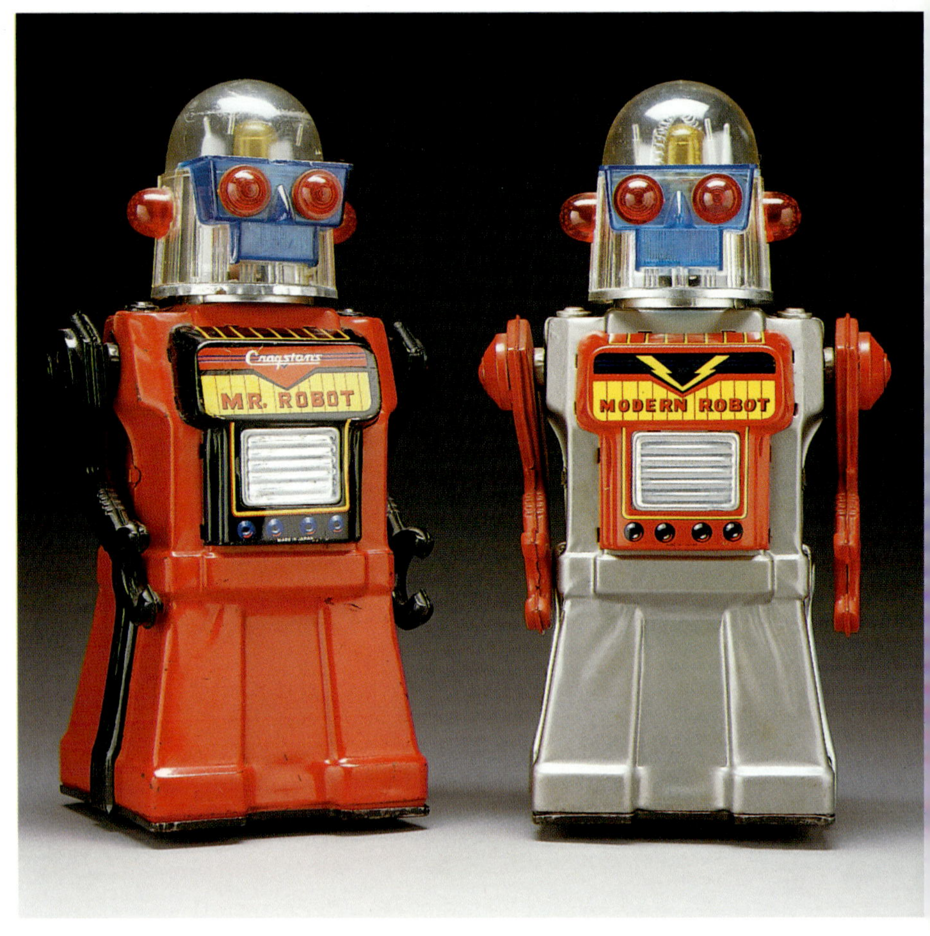

▲ 1950s, *Robots*, Yonezawa, 115 x 155 x 275 mm

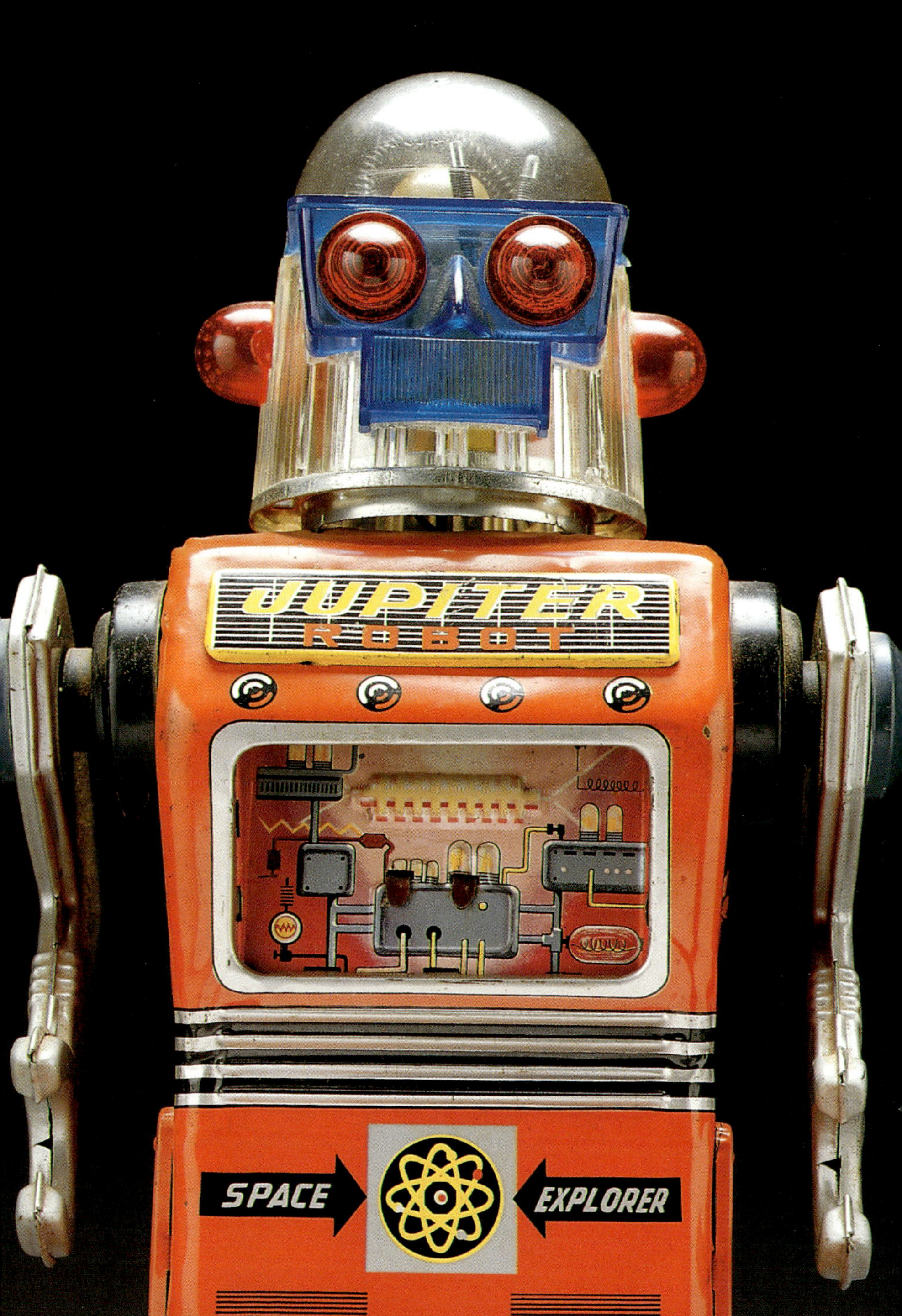
JUPITER
ROBOT
SPACE
EXPLORER

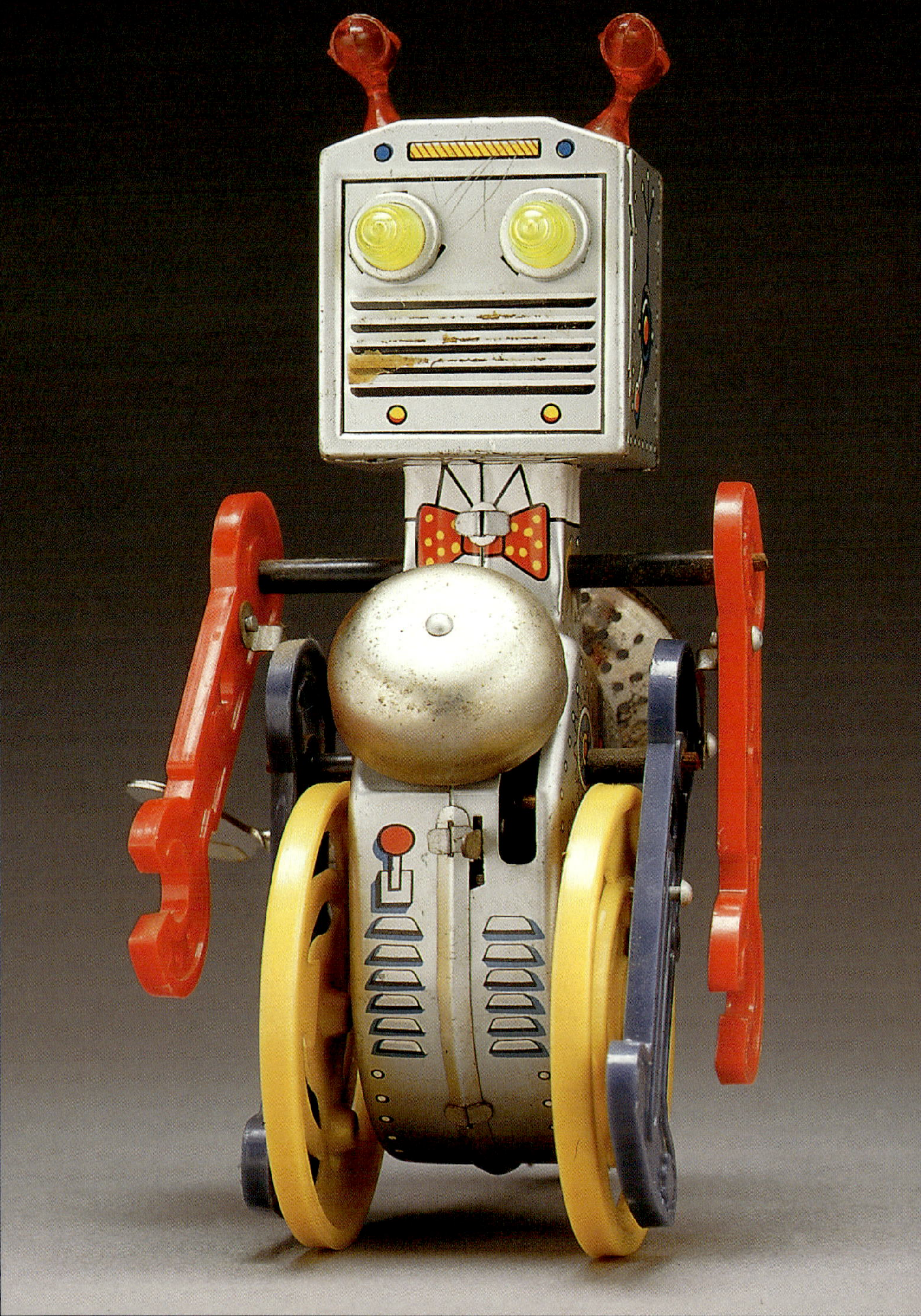

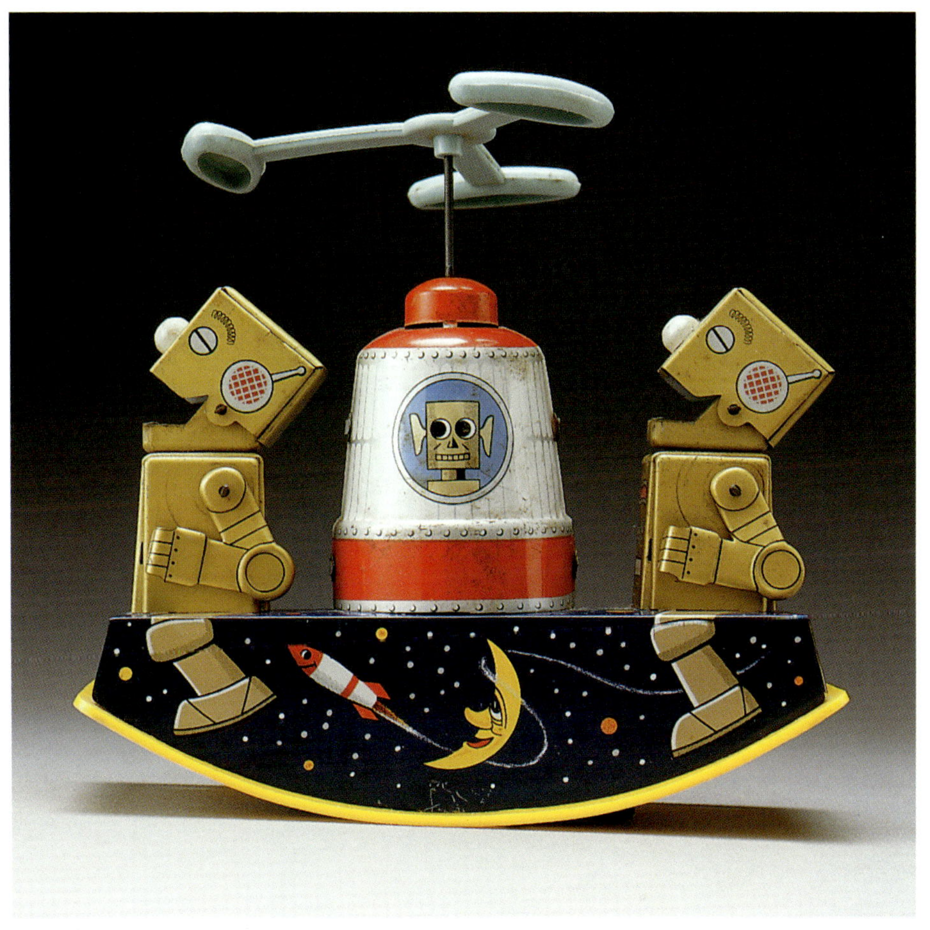

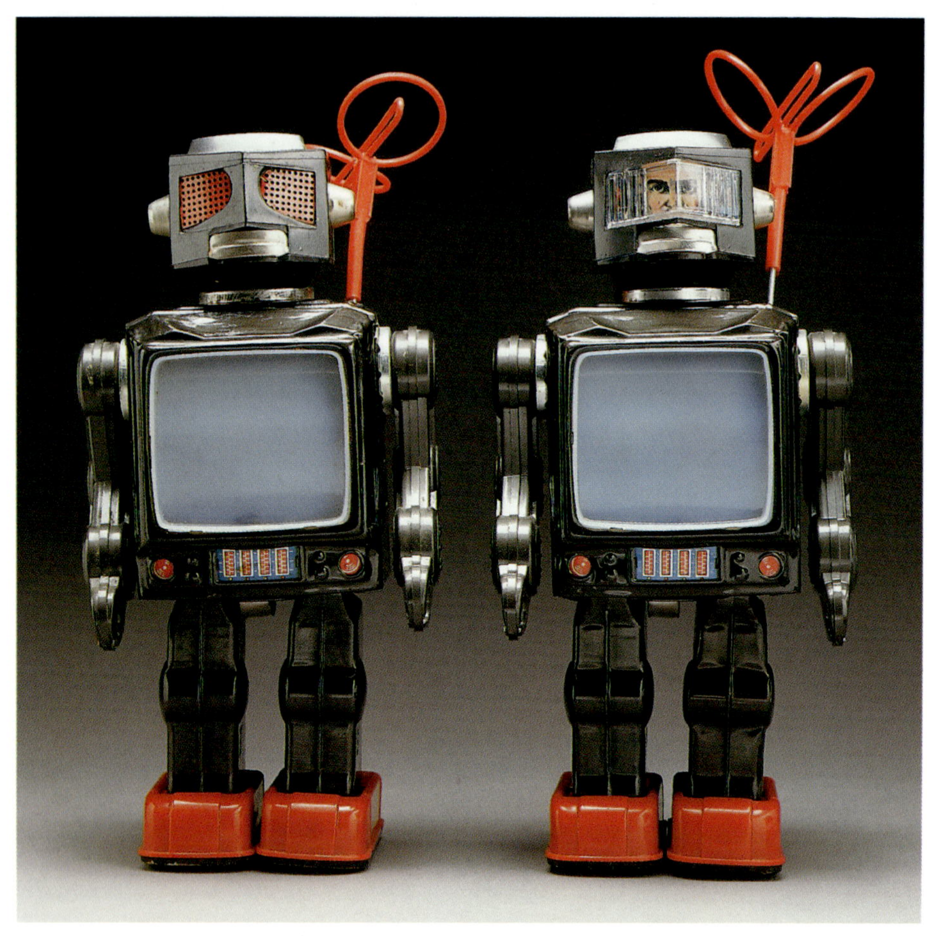

1960s, *TV Robots*, Horikawa, 110 x 136 x 325 mm

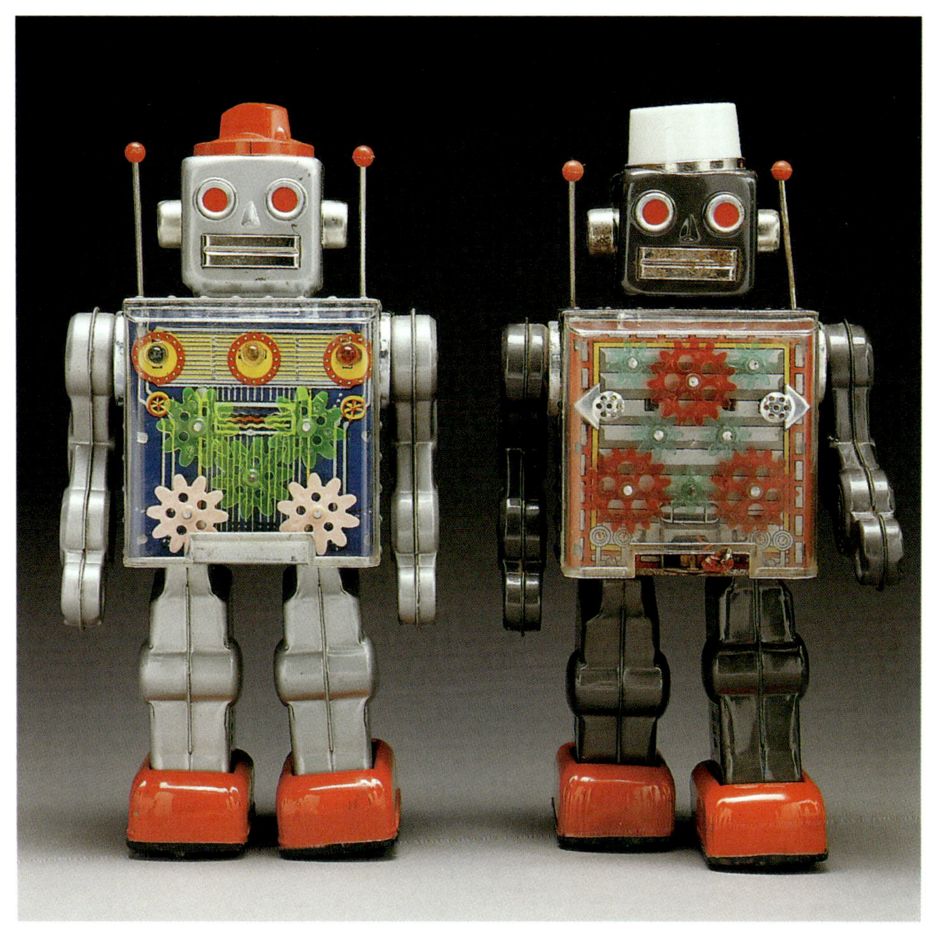

1950s, 1960s, *Gear Robots*, unknown, Horikawa, 82 x 138 x 282 mm

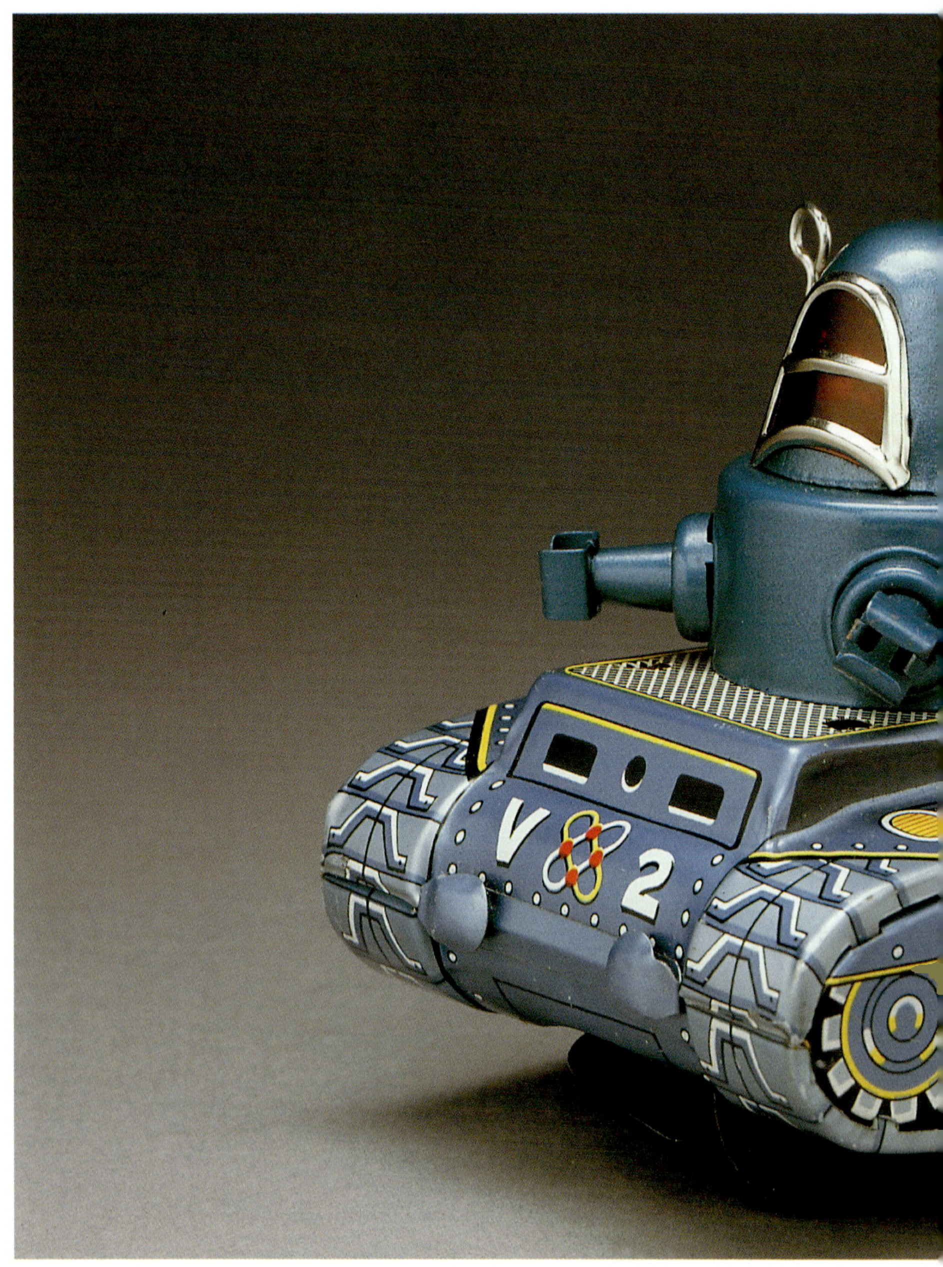

1950s, *Space Tank*, Yoshiya, 150 x 90 x 107 mm

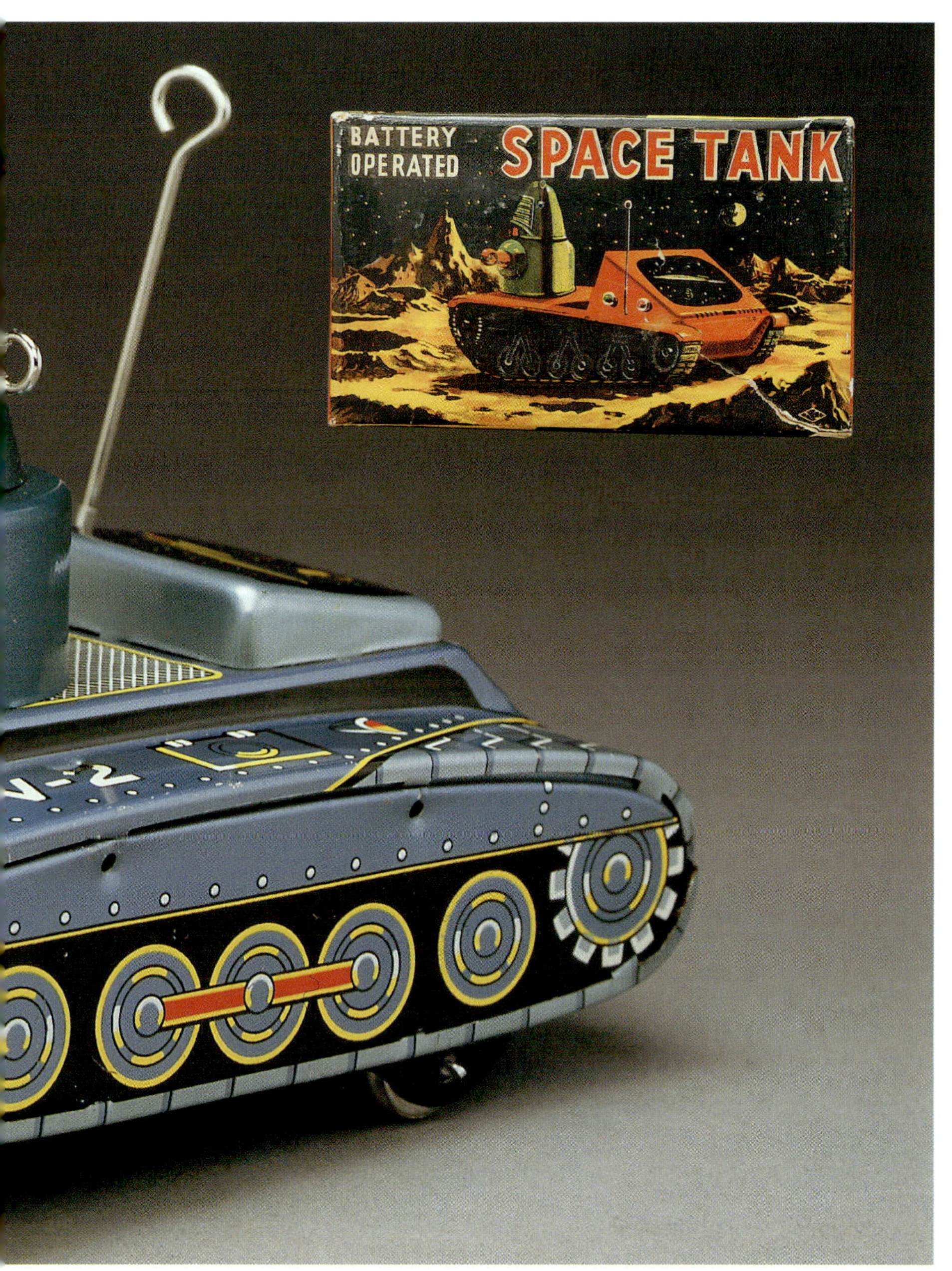

BATTERY OPERATED
SPACE TANK

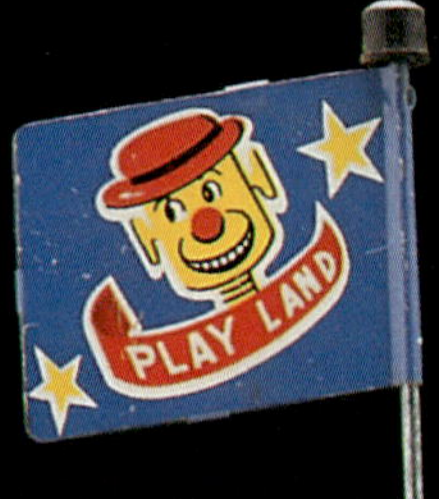
PLAY LAND
Driving Robot
Lucky

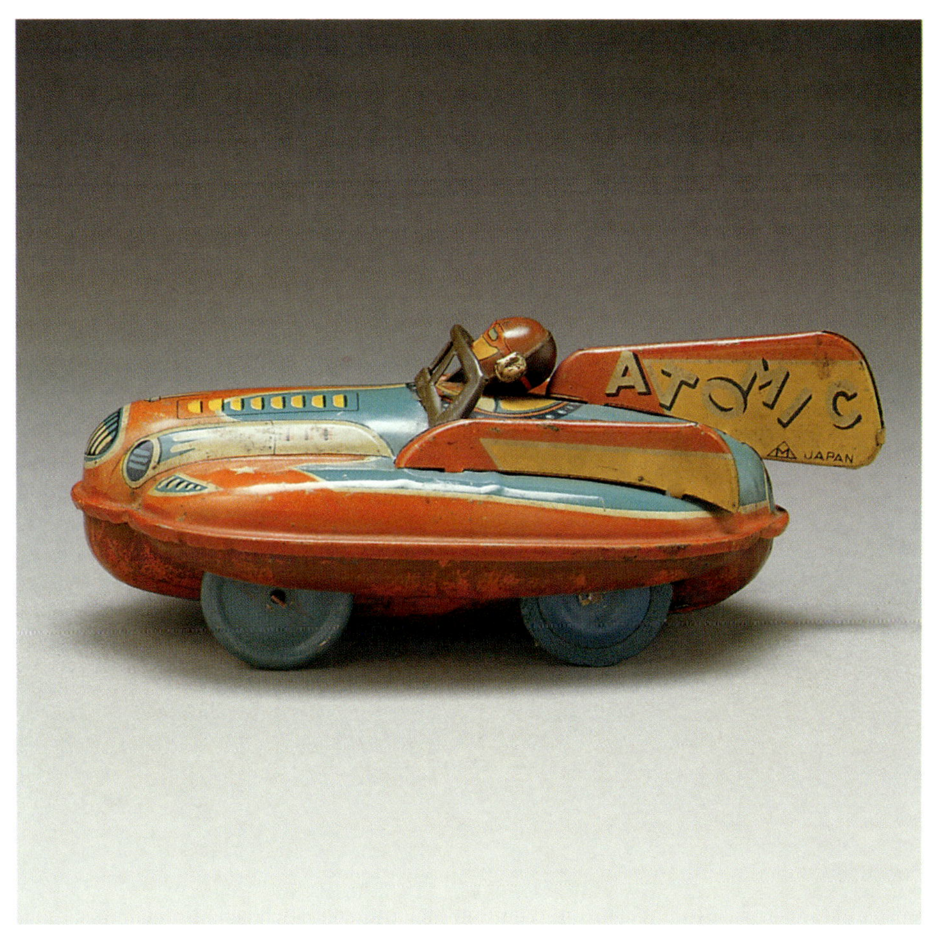

▲ 1950s, *Atomic X-8*, Mitsuhashi, 145 x 80 x 55 mm

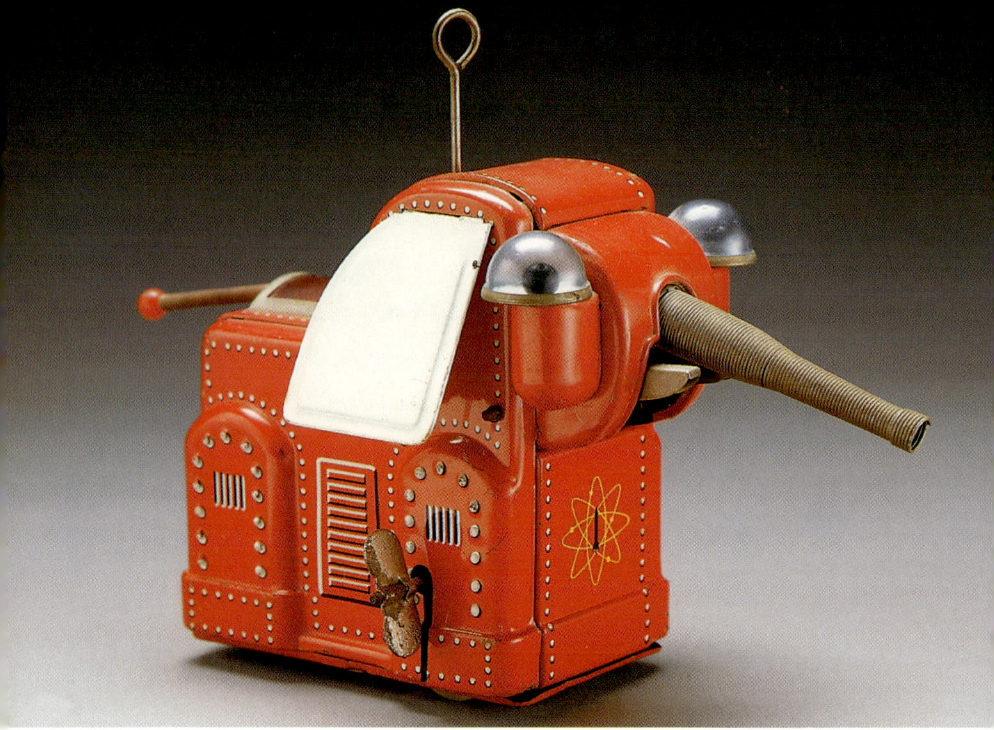

1950s, *Elephant Robot*, Yoshiya, 130 x 90 x 120 mm

1950s, *Space Dog*, Yoshiya, 185 x 75 x 115 mm

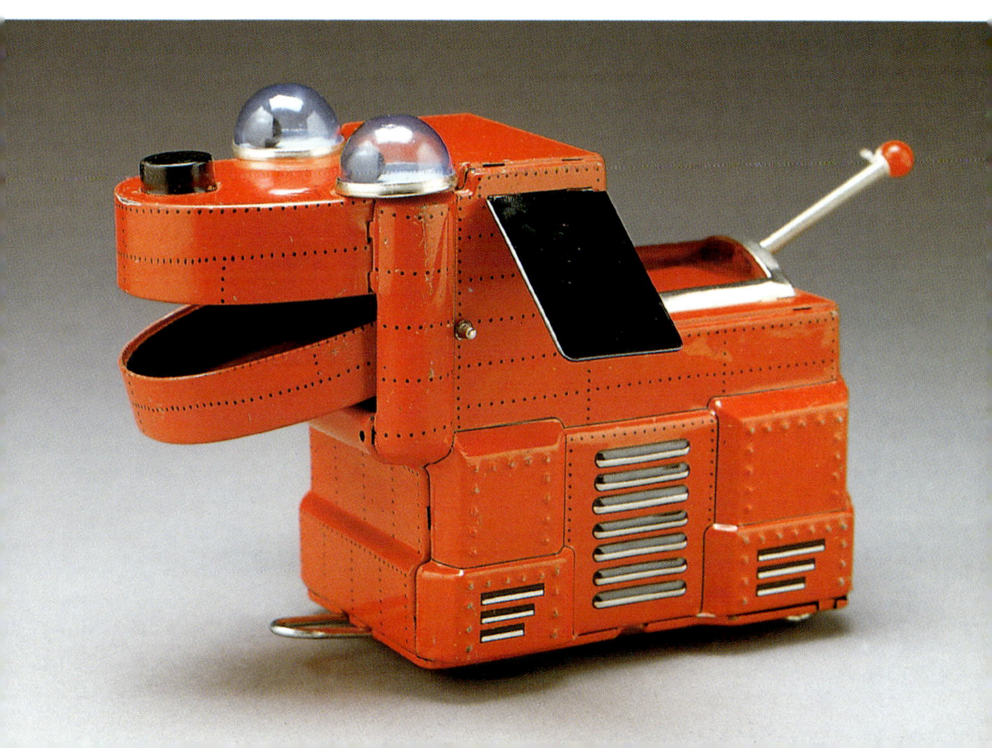

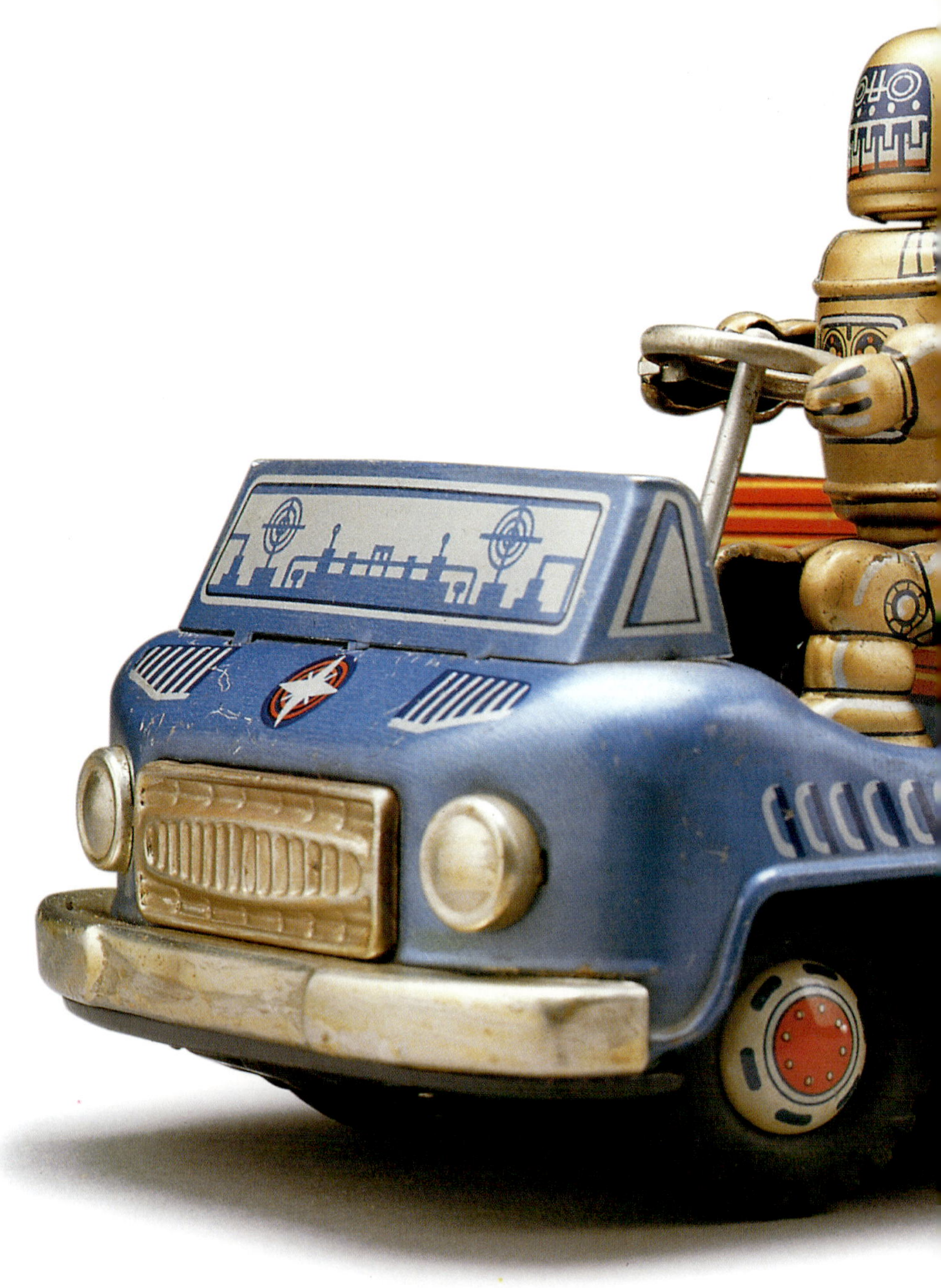

1950s, *Space Robot*, Yonezawa, 240 x 120 x 135 mm

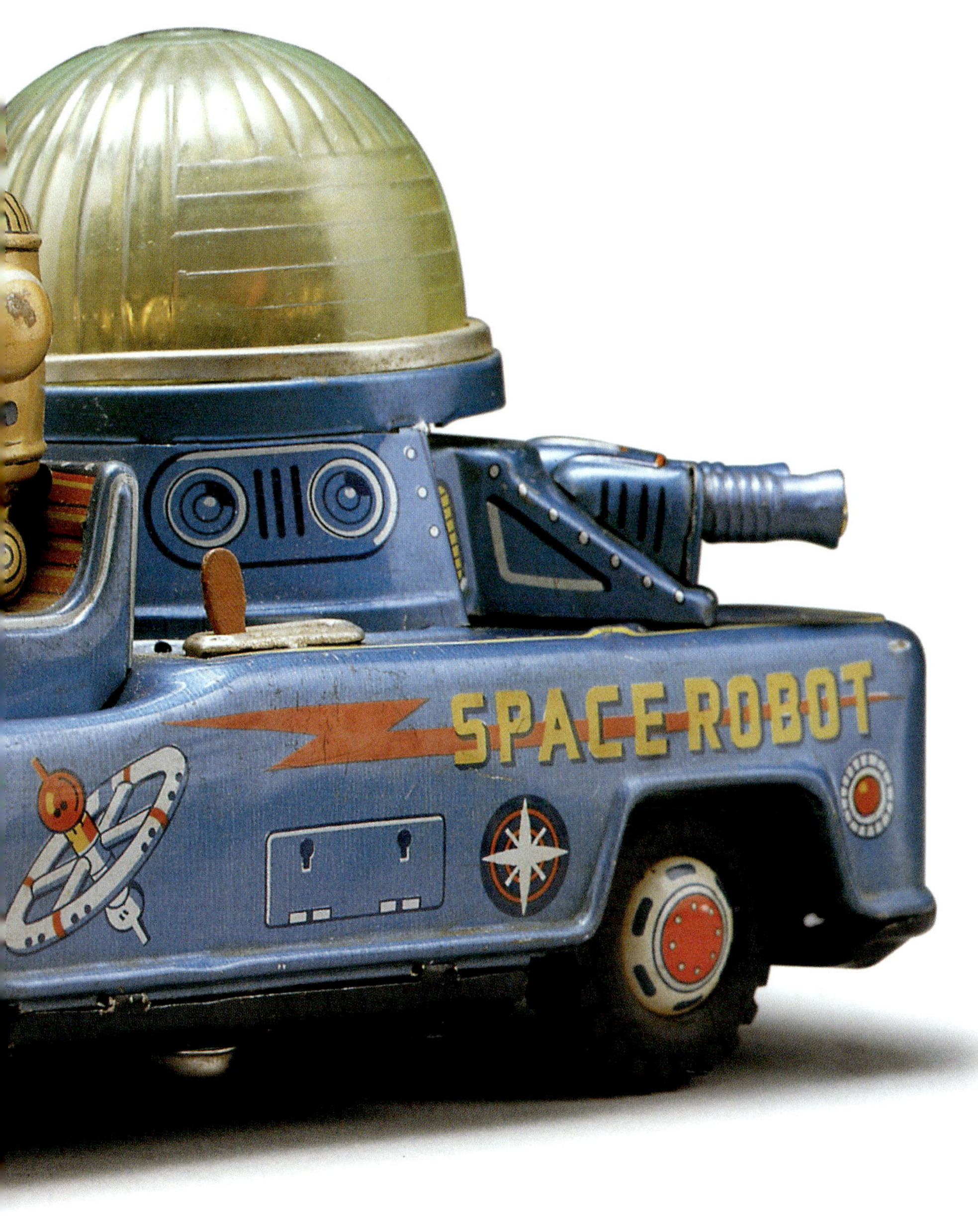

SPACE ROBOT

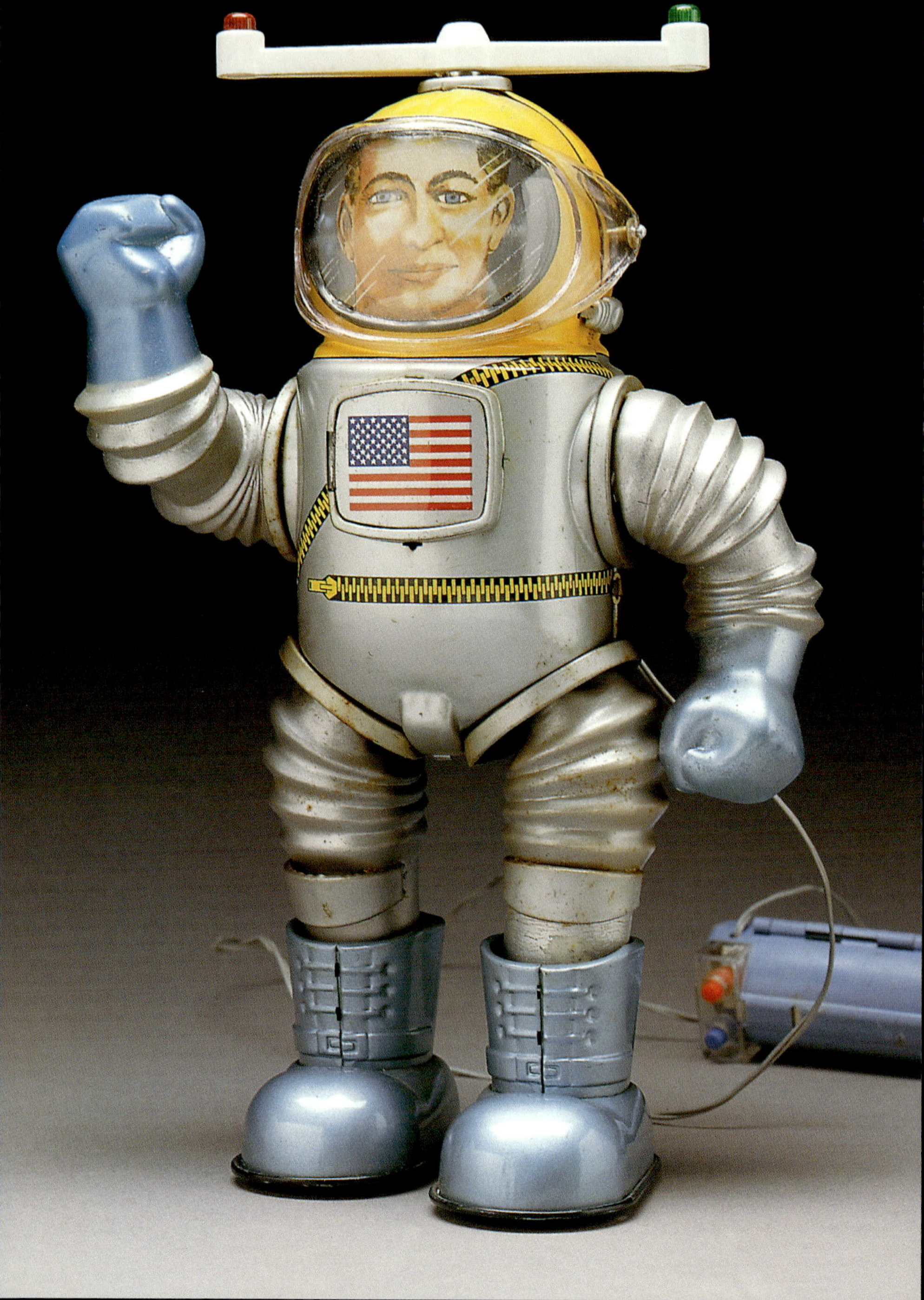

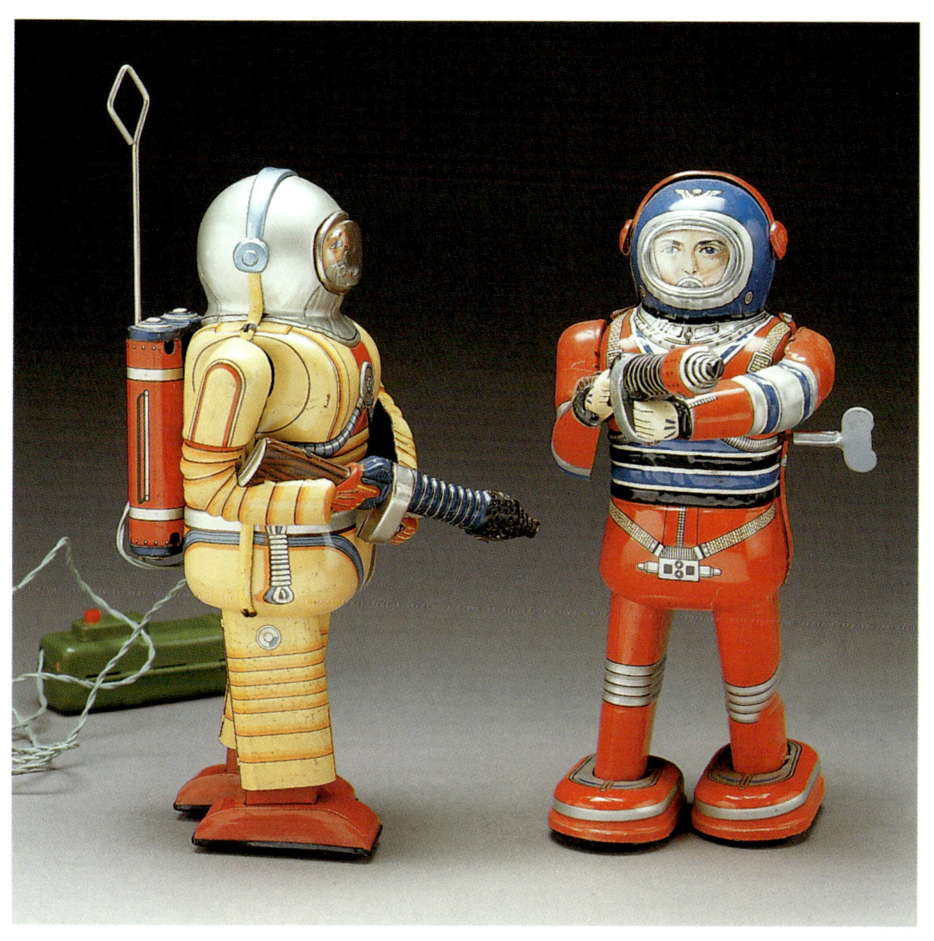

▲ 1950s, *Astronauts*, Nomura, Daiya, 75 x 95 x 230 mm

1950s, *Giant Robot*, Horikawa, 148 x 220 x 400 mm ▶

▲ 1960s, *Space Man*, *Space Commander*, Horikawa, 95 x 140 x 280, 120 x 140 x 260 mm

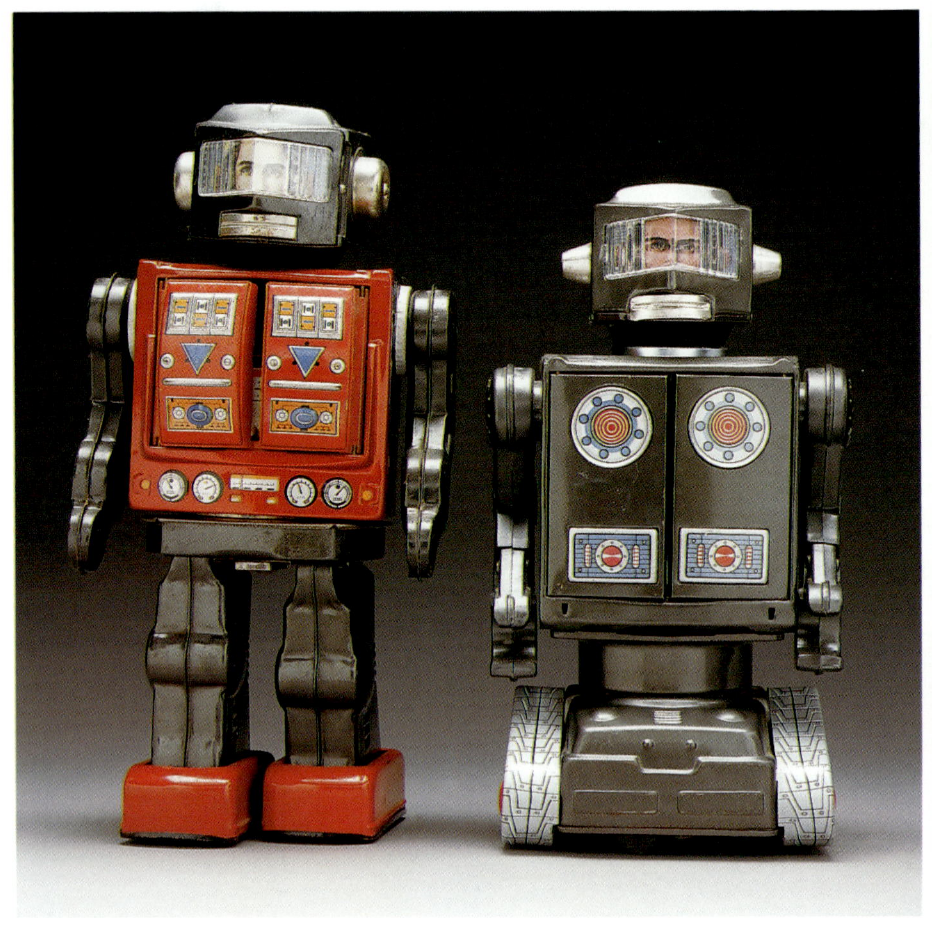

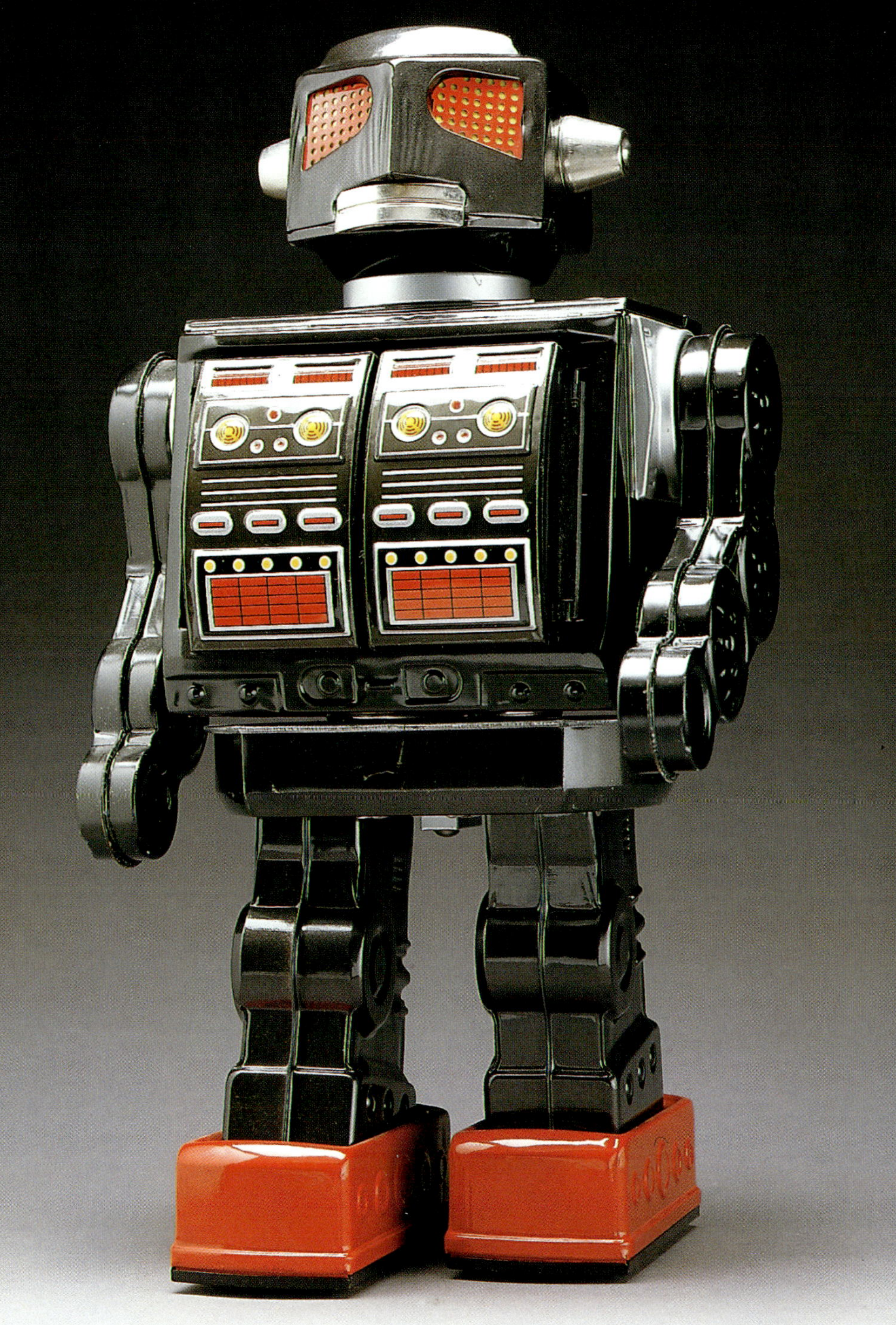

1950s, *Robby*, Yoshiya, 65 x 92 x 160 mm

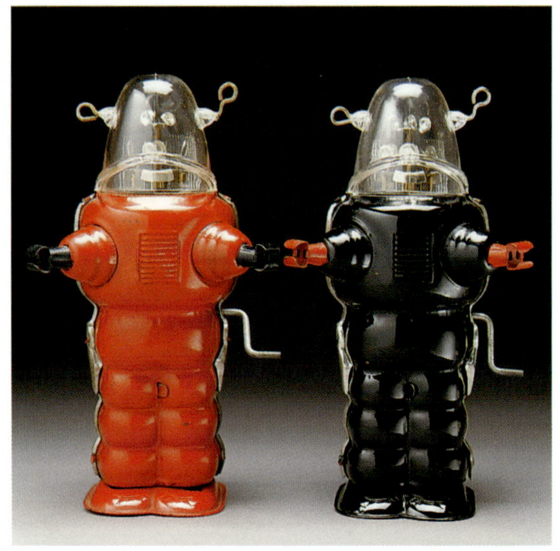

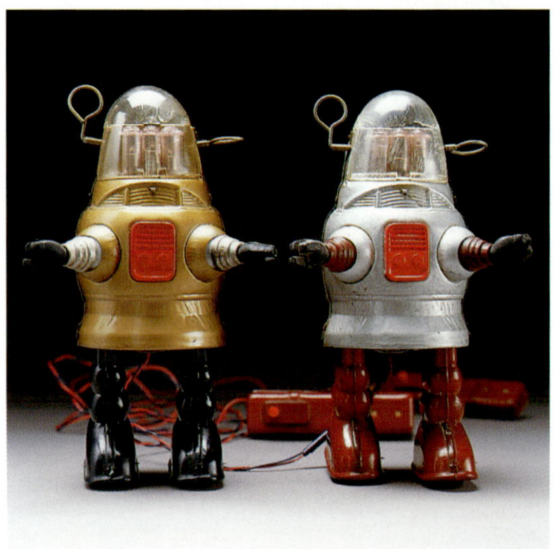

1950s, *Robby*, Nomura, 95 x 125 x 215 mm

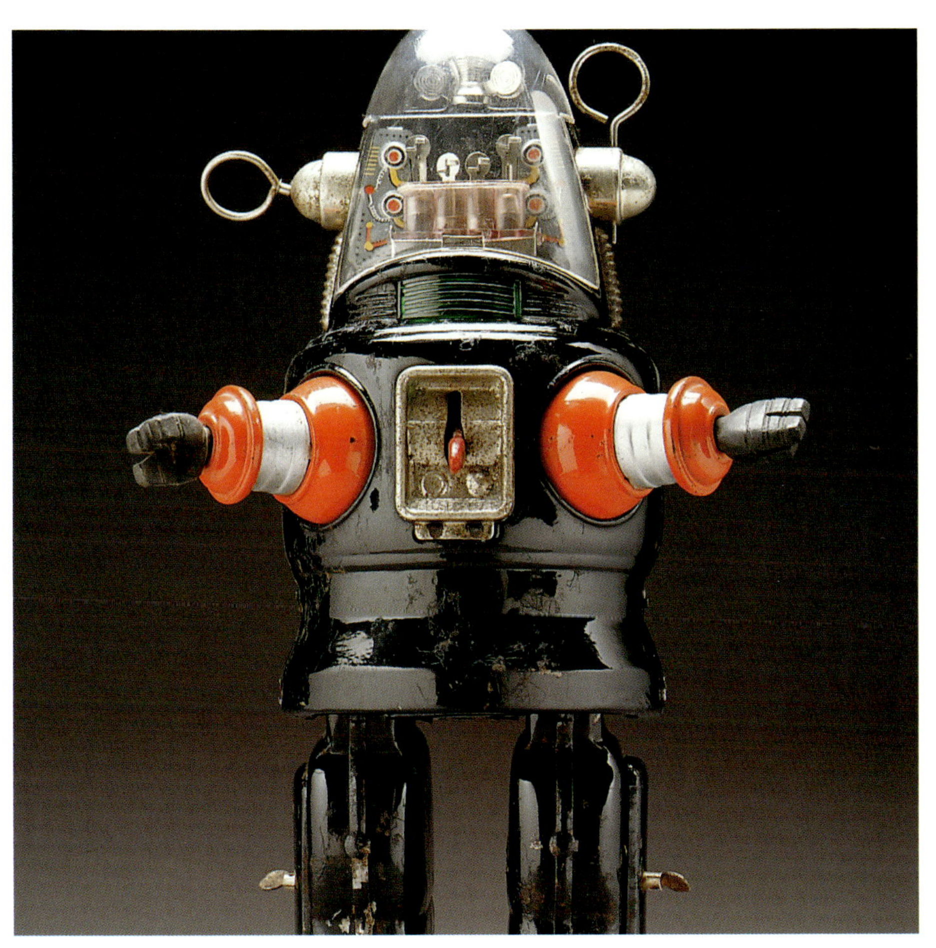

1950s, *Robby*, Nomura, 135 x 175 x 310 mm

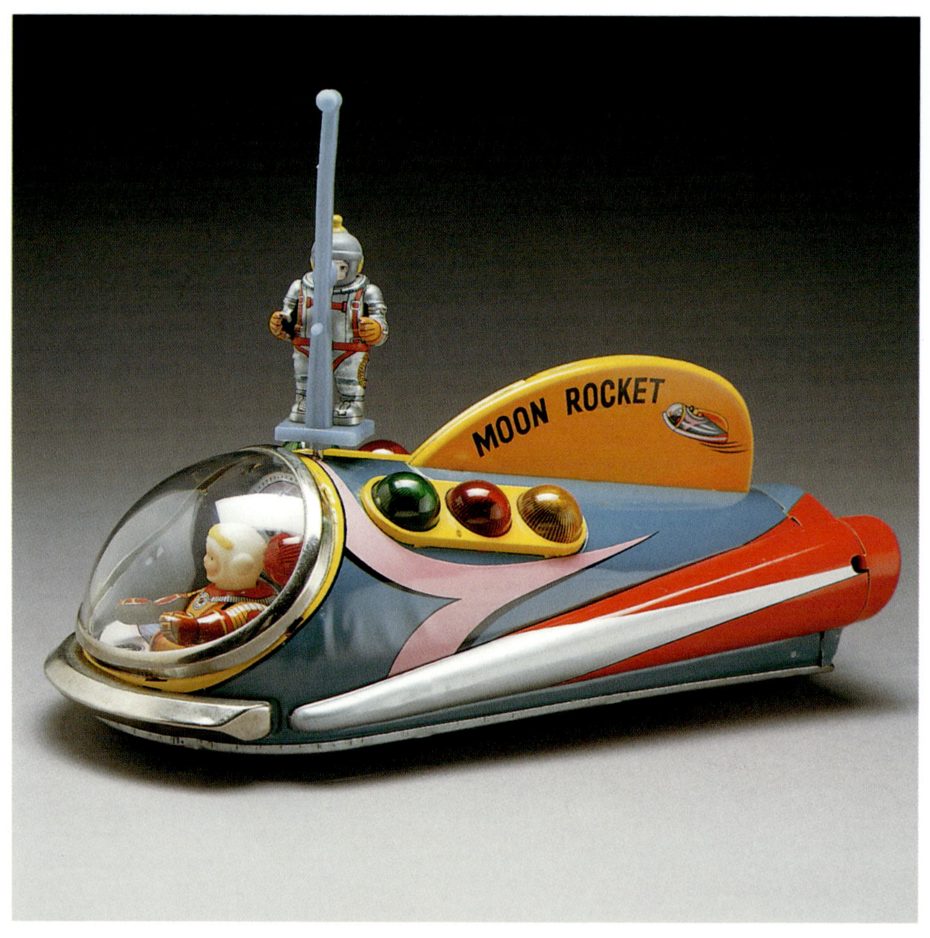

1960s, *Moon Rocket*, Masudaya, 235 x 112 x 170 mm

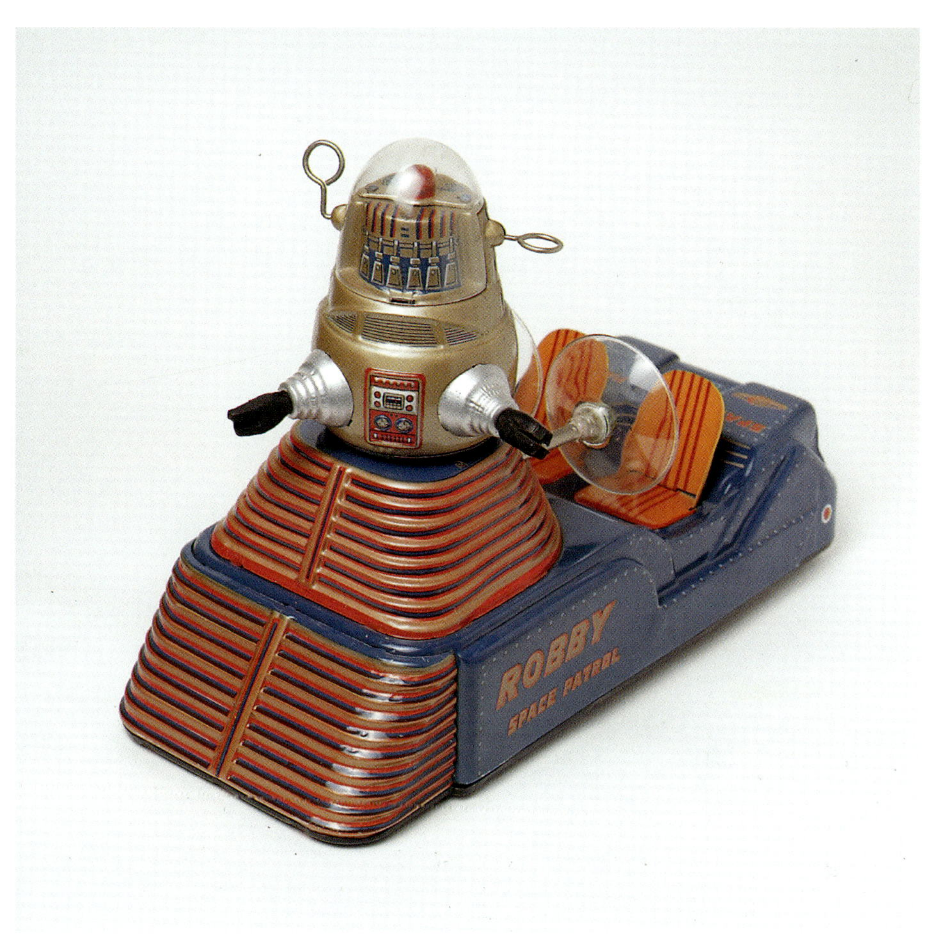

1950s, *Robby Space Patrol*, Nomura, 330 x 150 x 225 mm

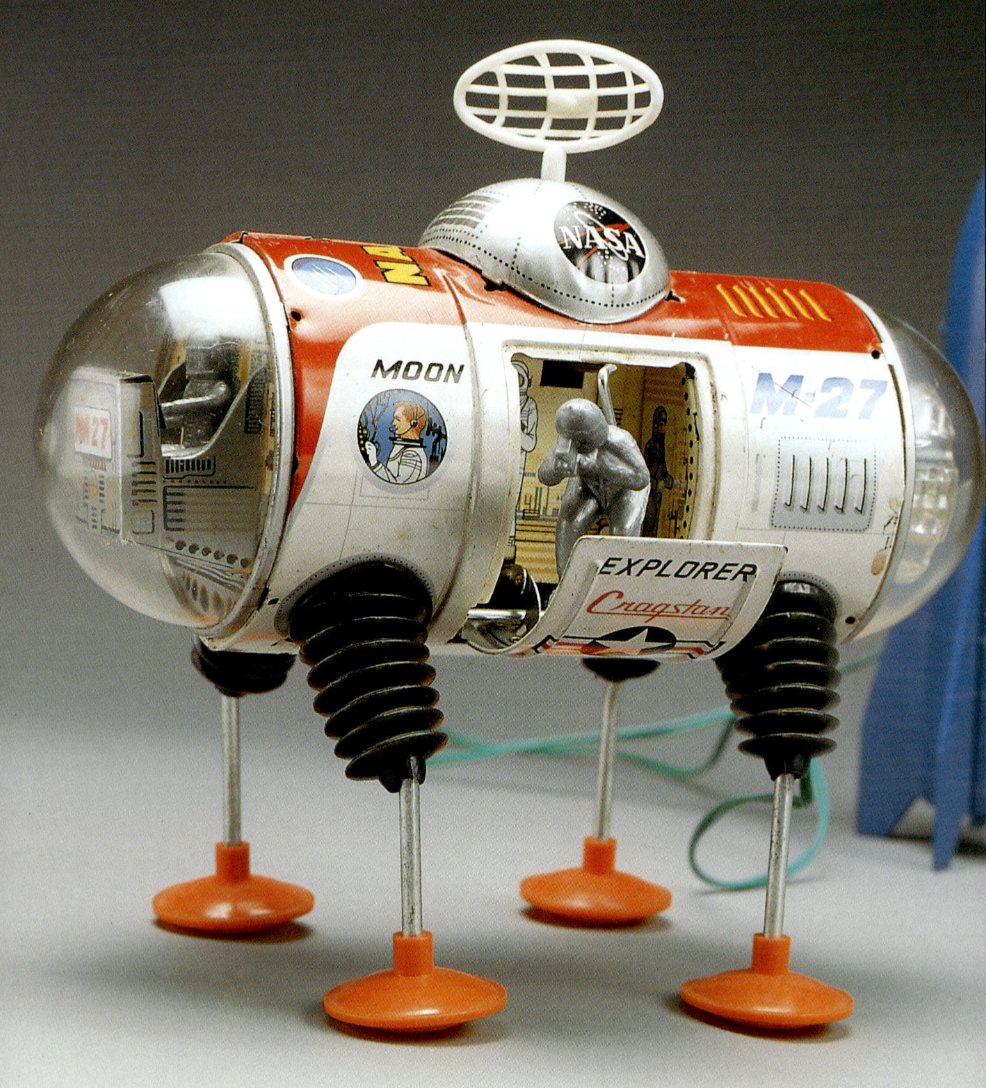

NASA
MOON
M-27
EXPLORER
Cragstan

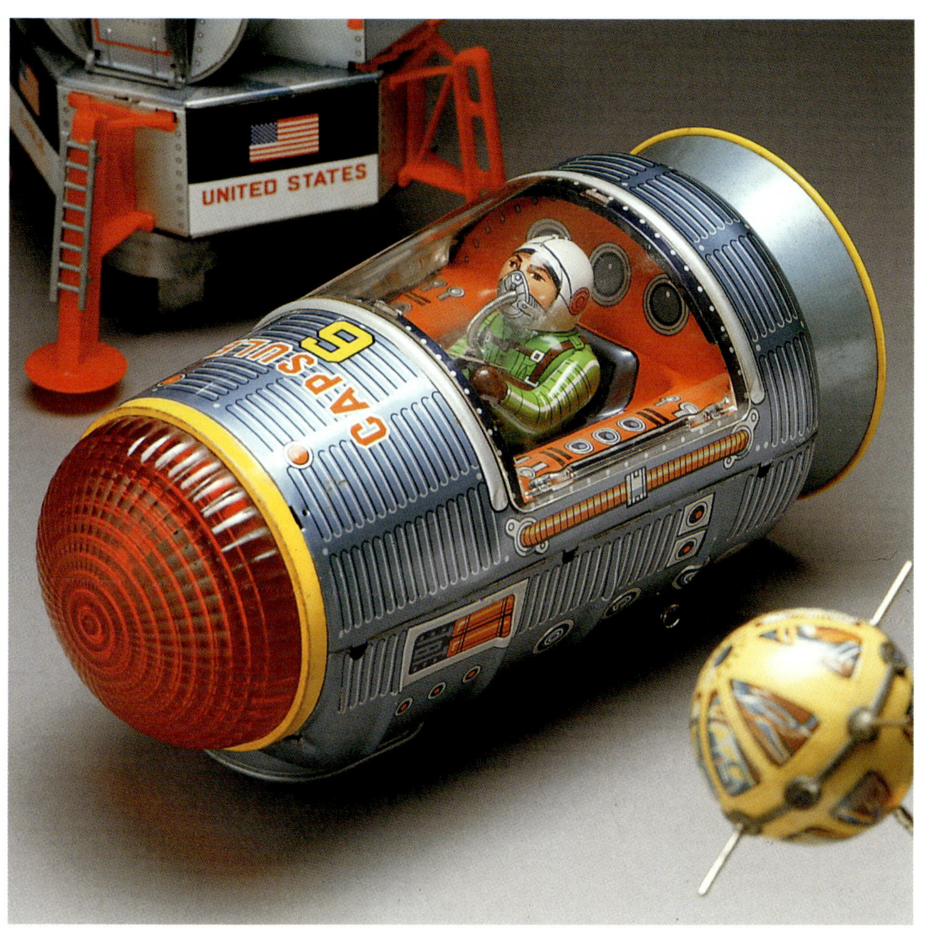

▲ 1950s, *Capsule 6*, Masudaya, 255 x 120 x 130 mm

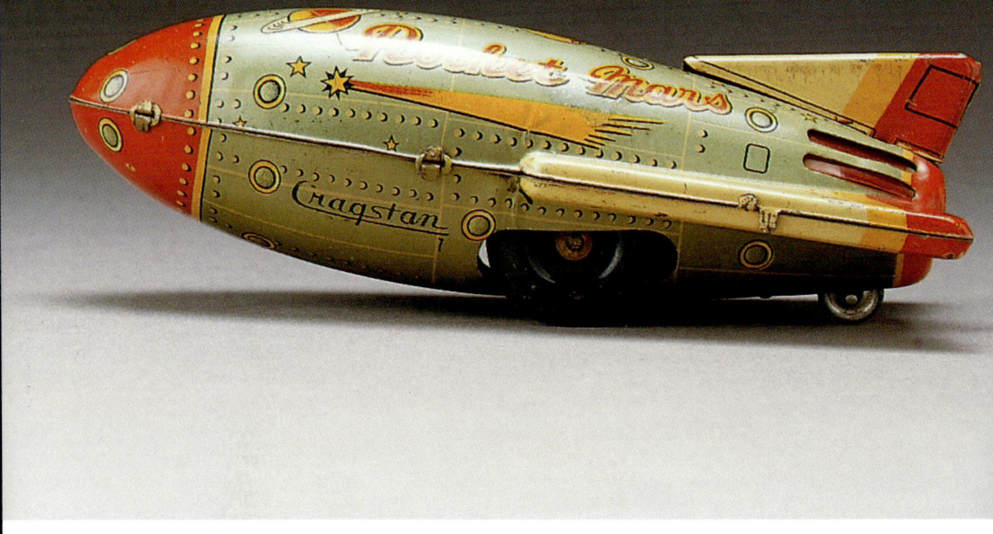

1950s, *Rocket Mars*, Cragstan, 140 x 62 x 47 mm

1950s, *Rocket 54*, unknown, 168 x 58 x 45 mm

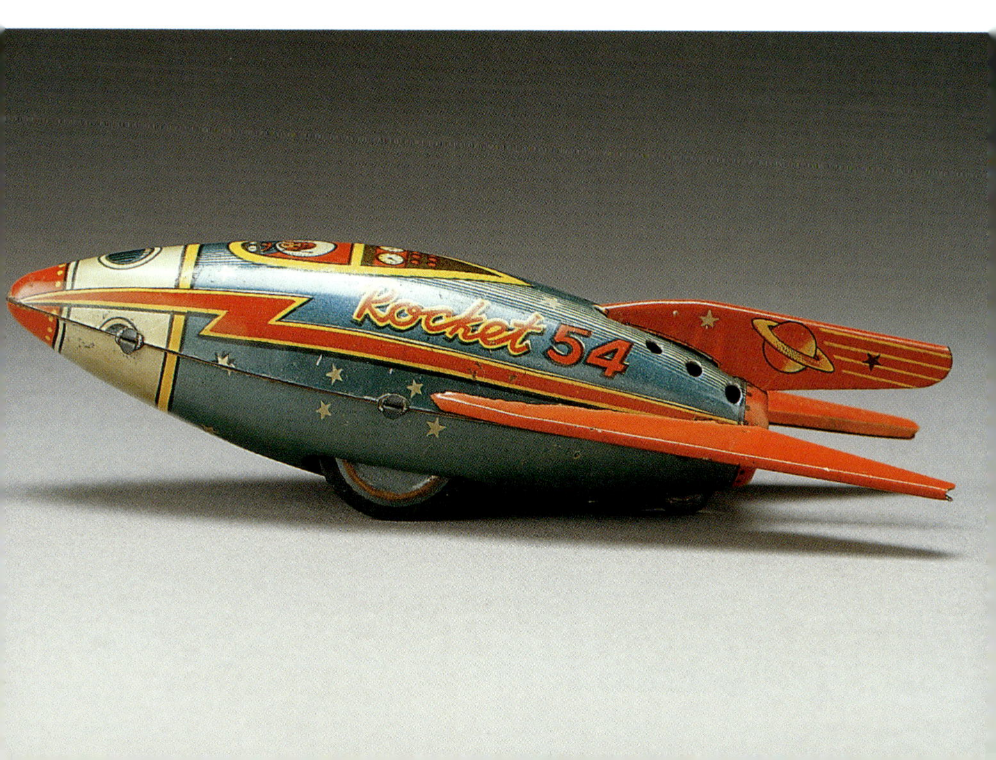

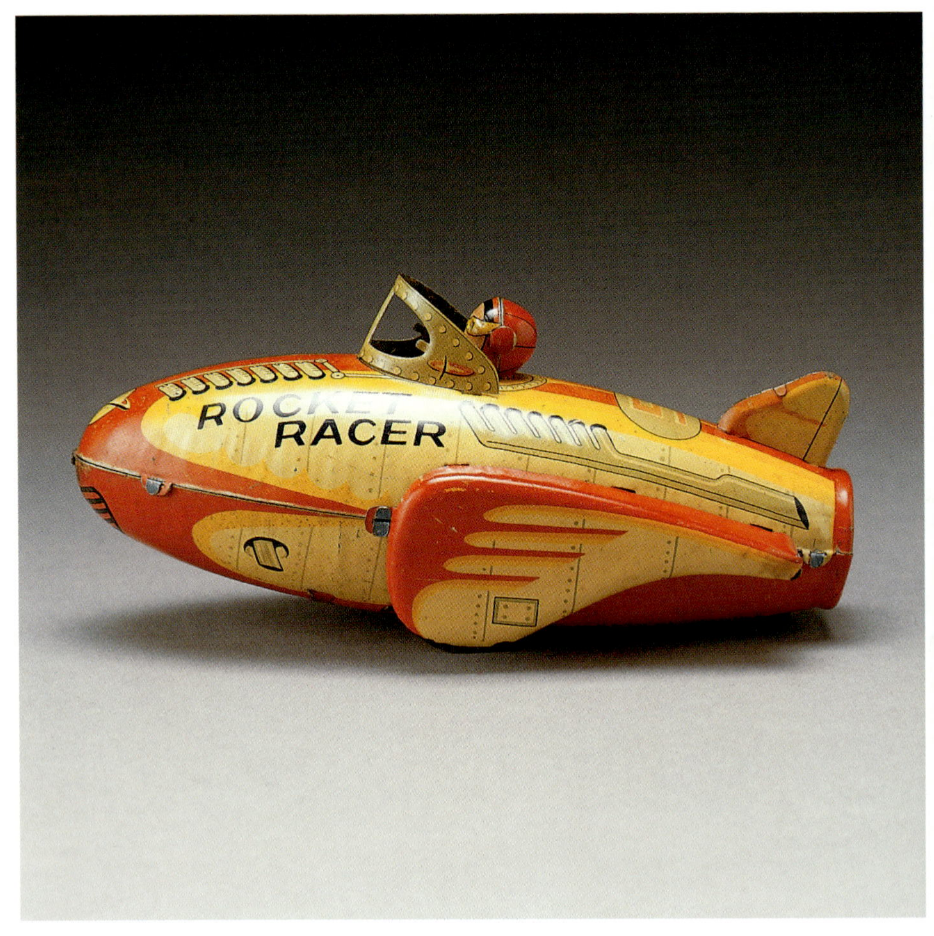

▲ 1950s, *Rocket Racer*, Masudaya, 165 x 70 x 75 mm

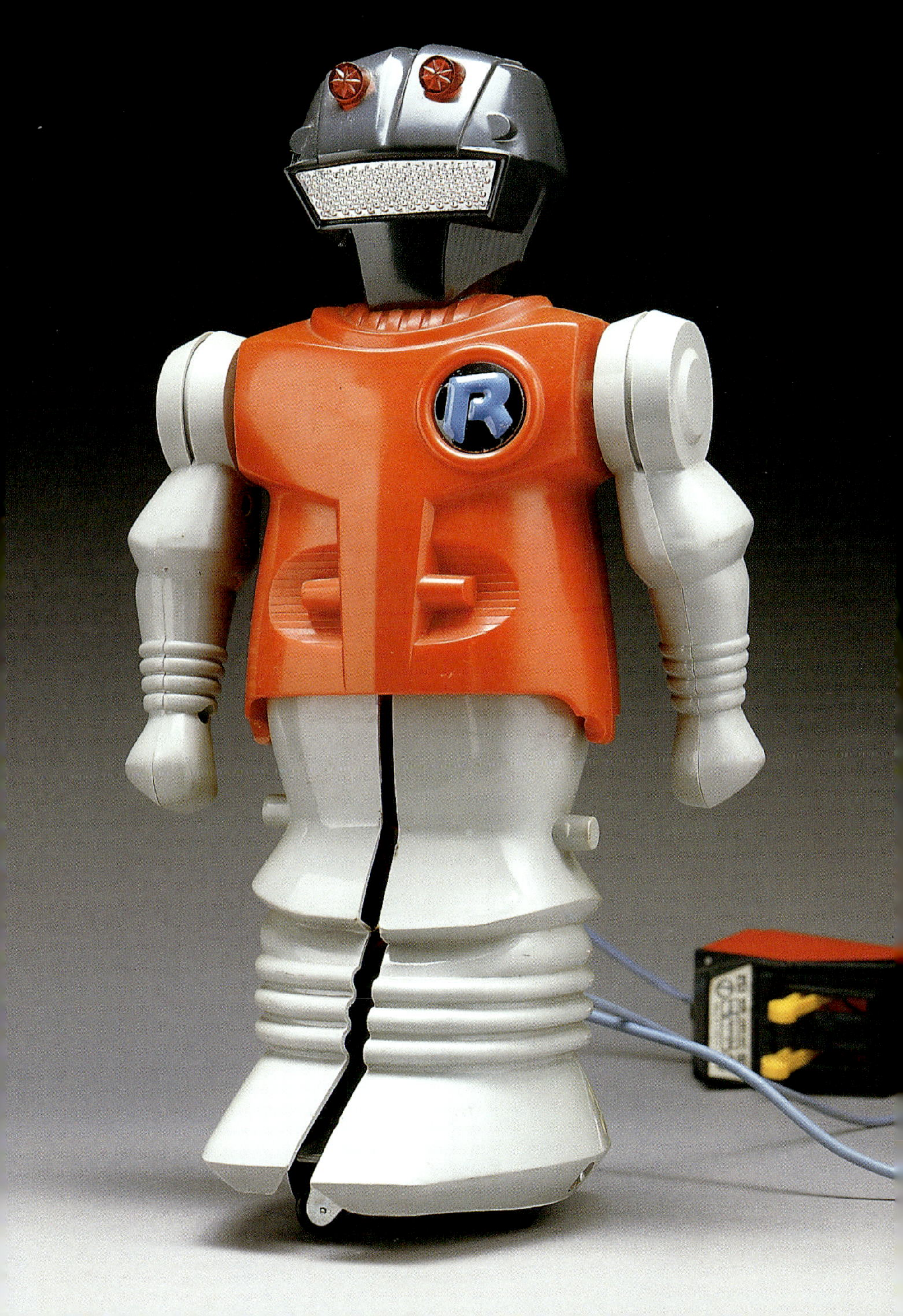

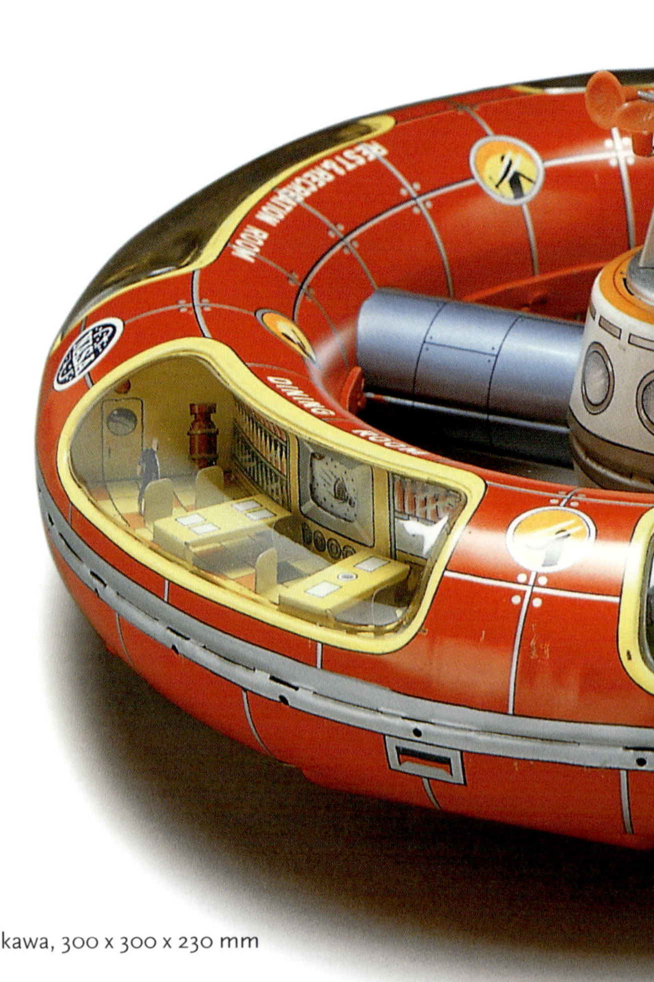

1950s, *Space Station*, Horikawa, 300 x 300 x 230 mm

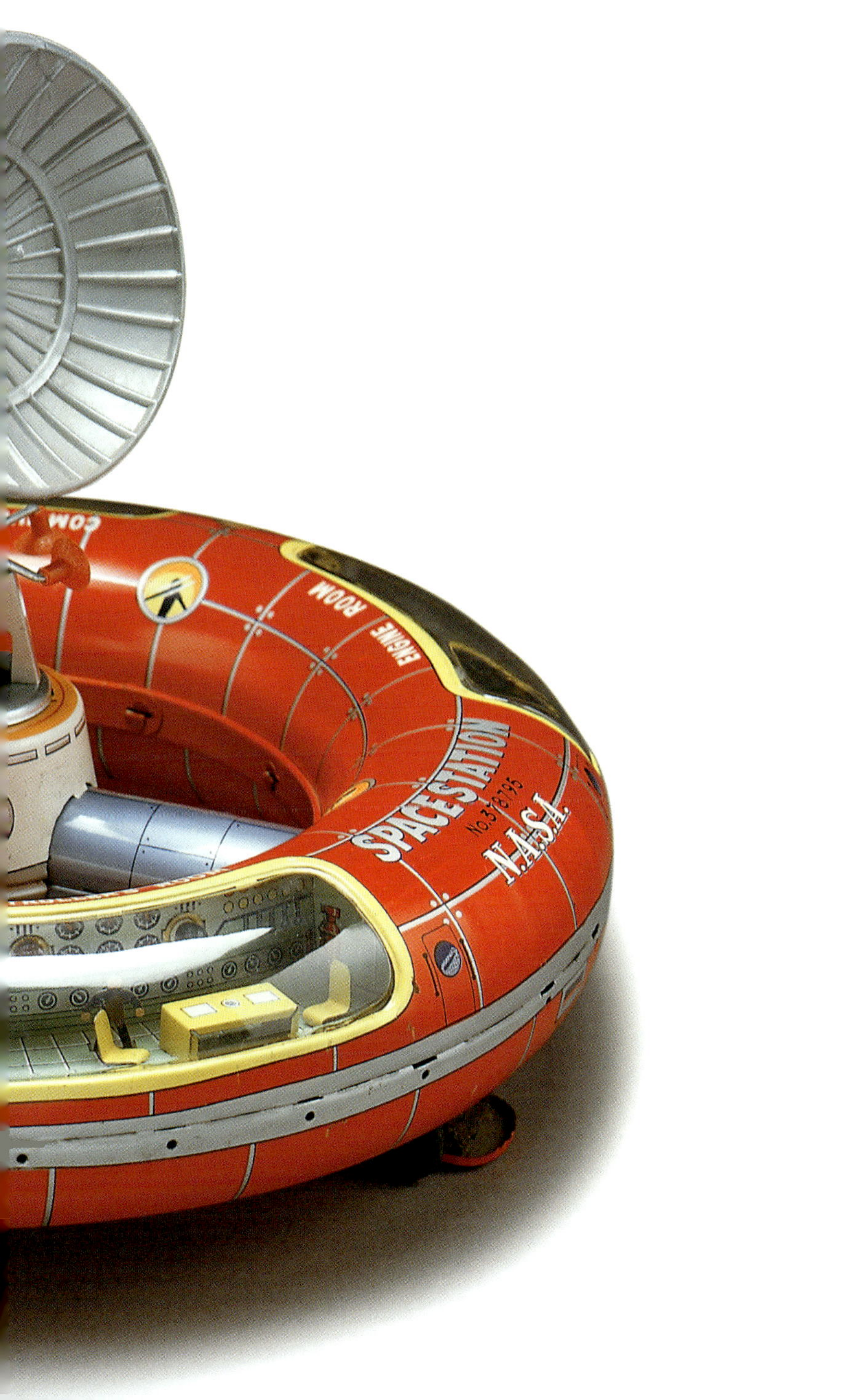

ENGINE ROOM
SPACE STATION
No.3/8/35
N.A.S.A.

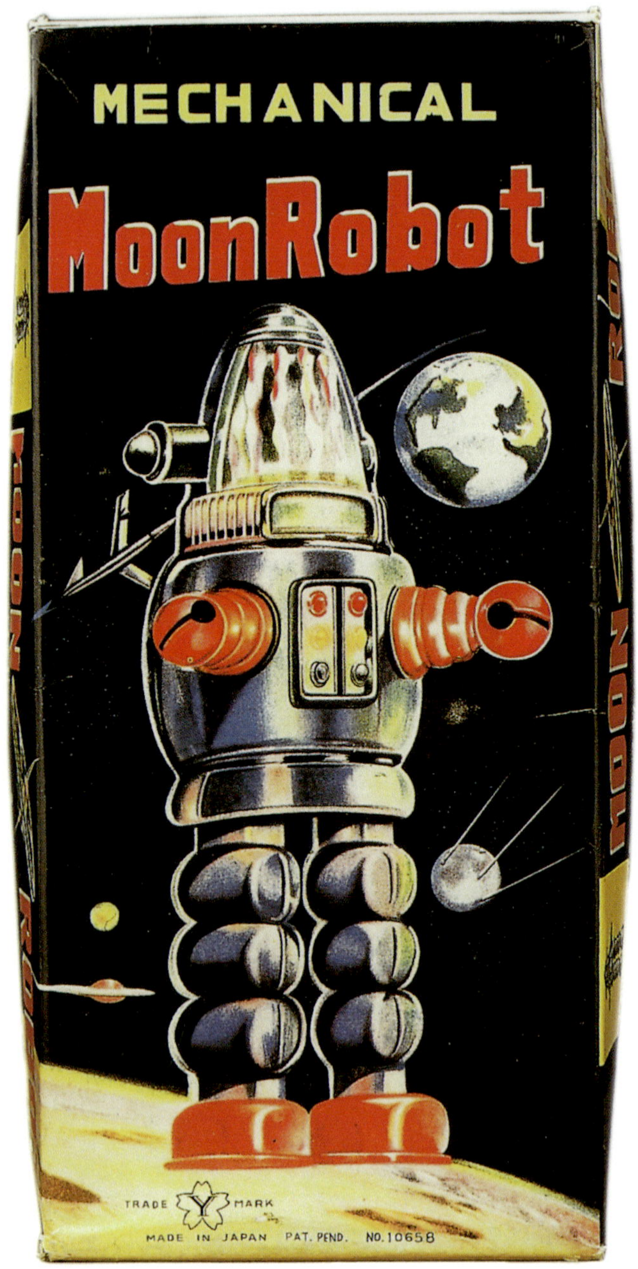

1950s, *Moon Robot*, Yonezawa, 110 x 125 x 260 mm

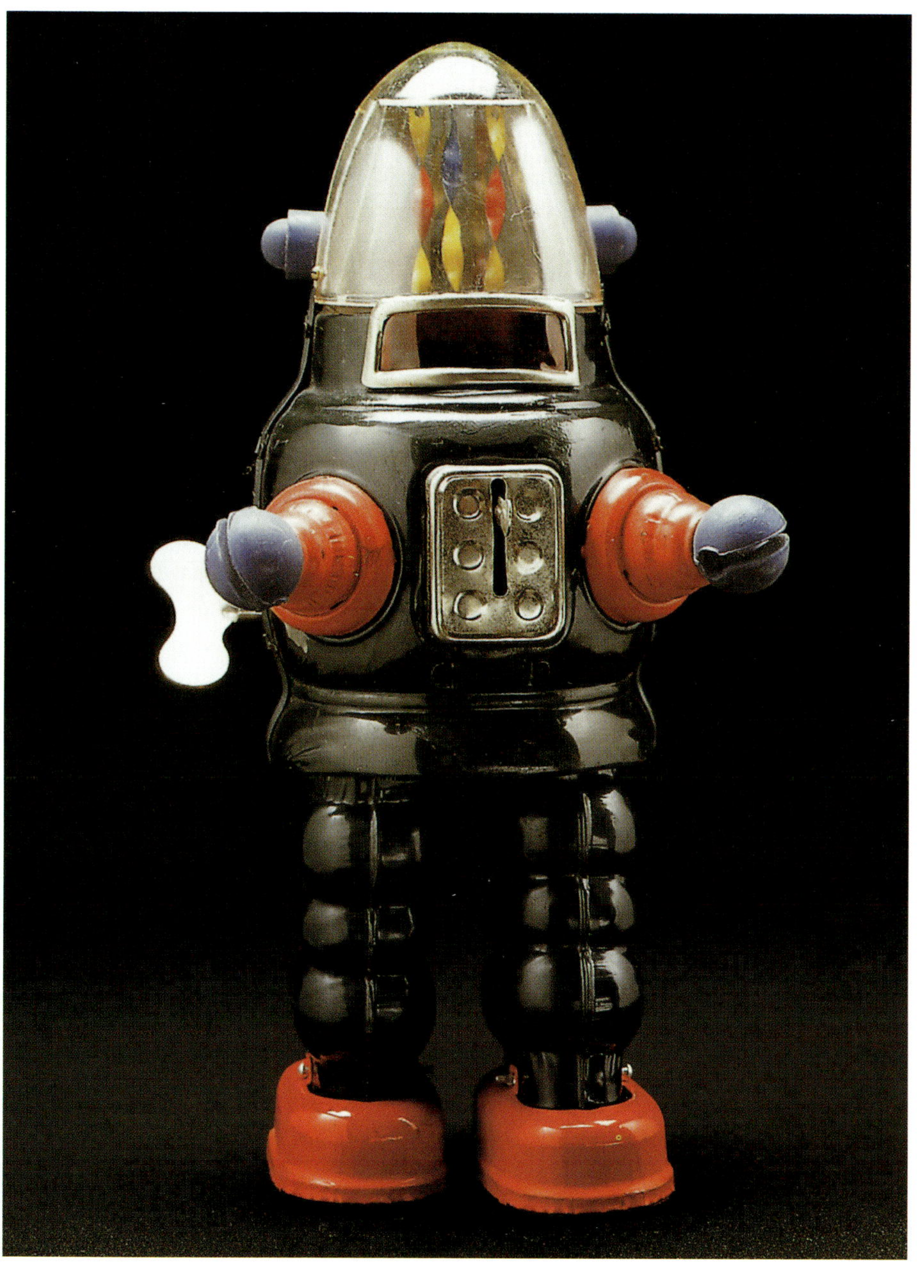

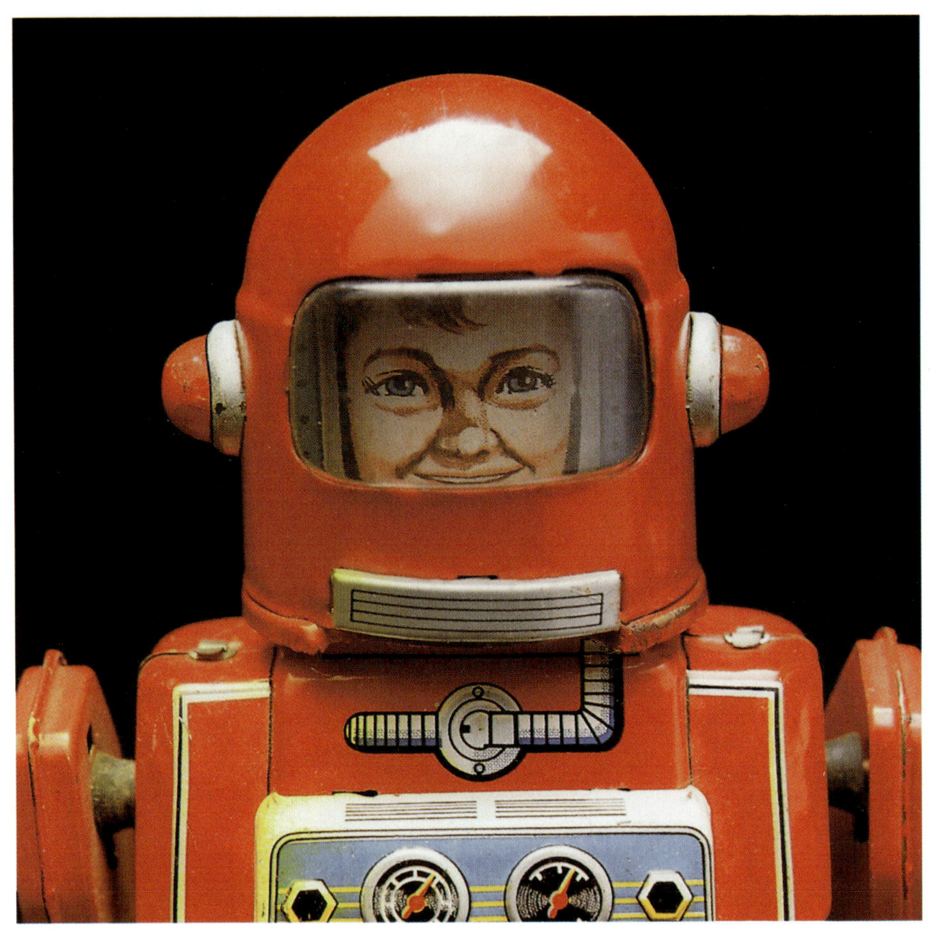

1950s, *Astronaut*, Yonezawa, 95 x 125 x 240 mm

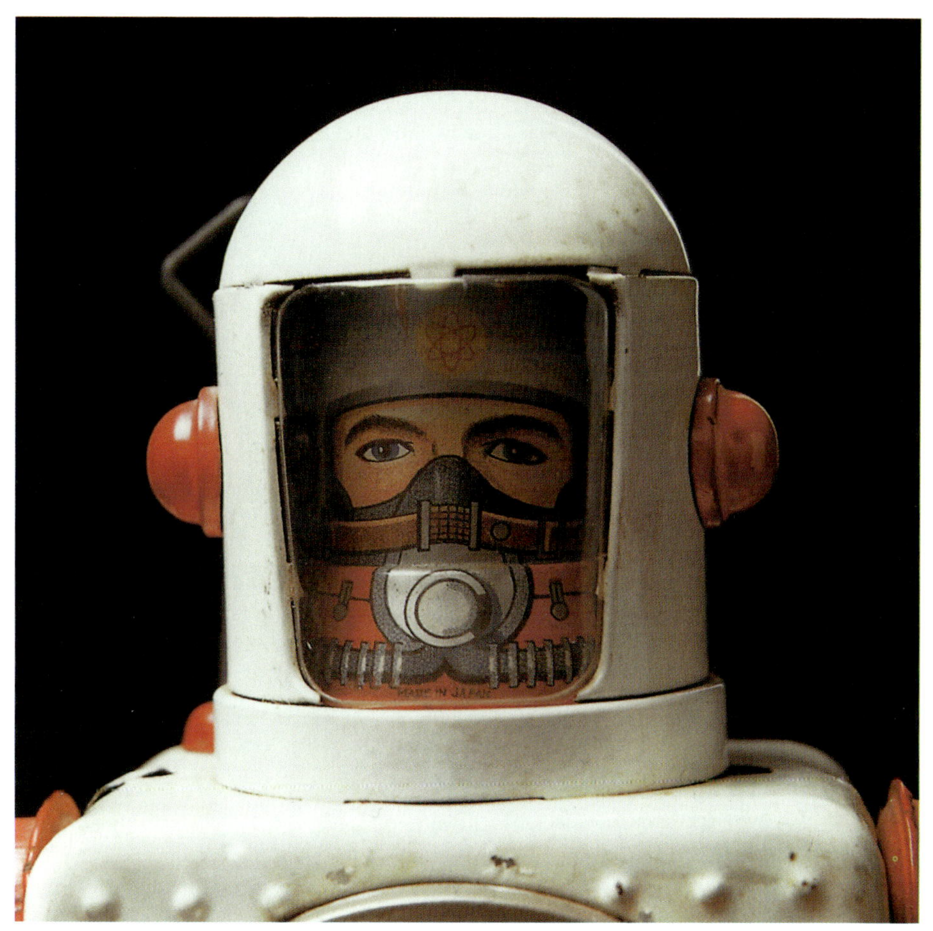

1950s, *Space Scout*, Yonezawa, 100 x 115 x 240 mm

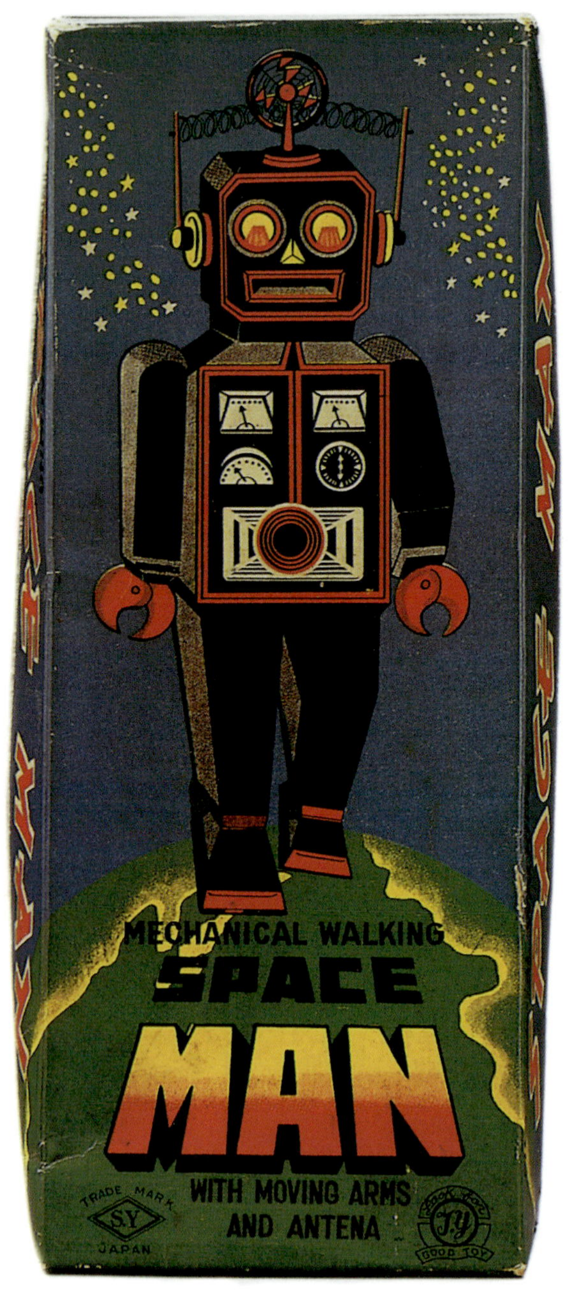

1950s, *Space Man*, Yoshiya, 65 x 75 x 215 mm

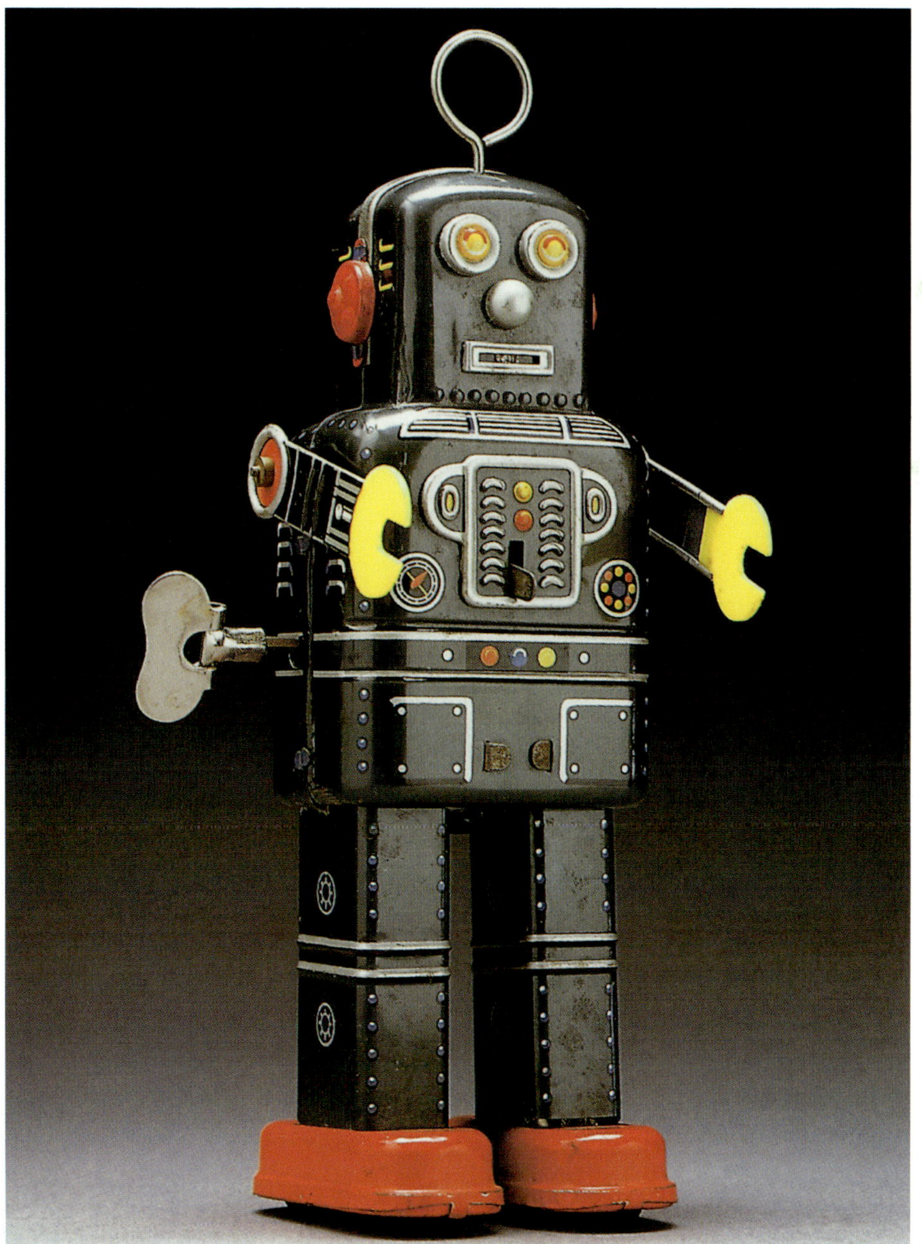

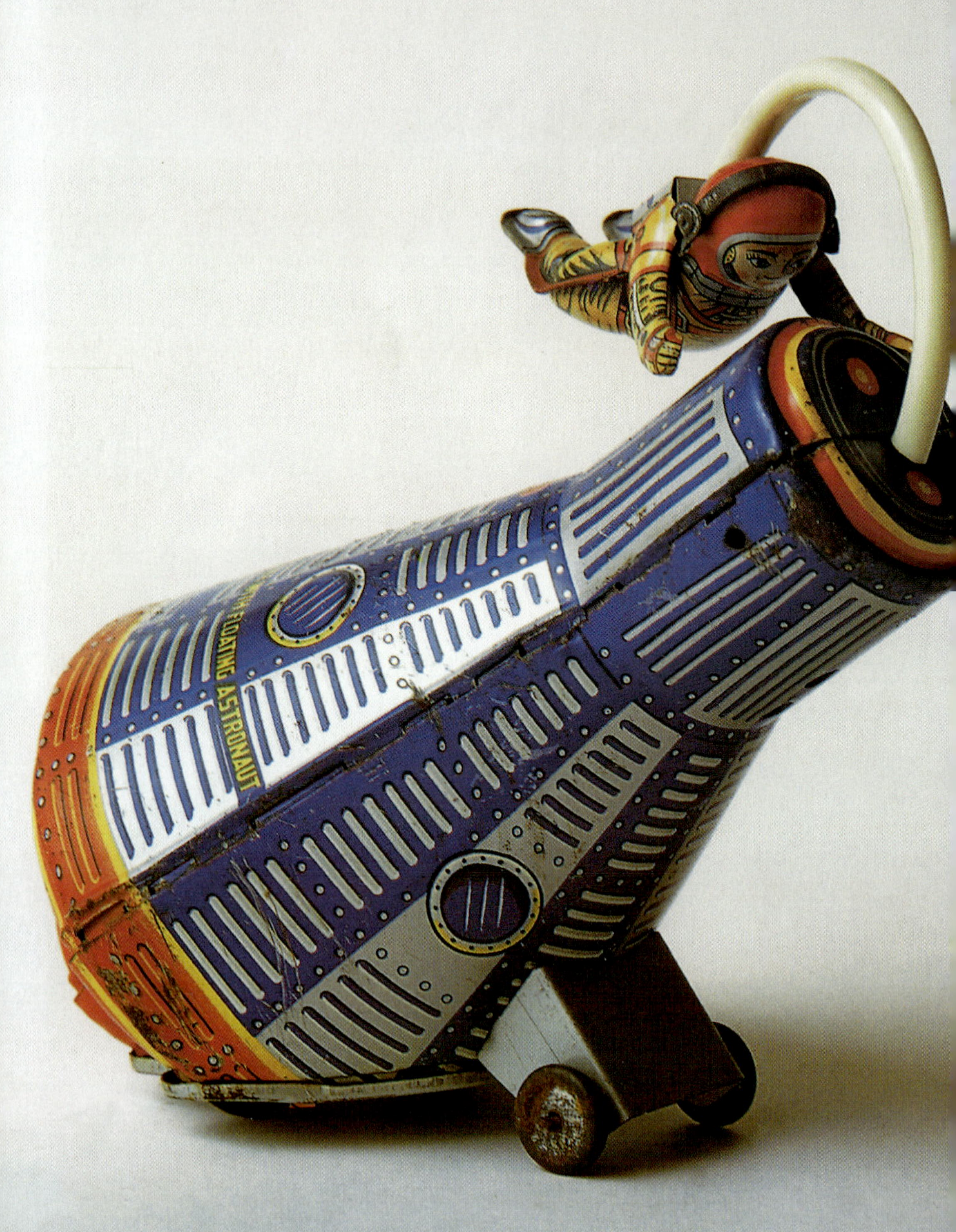
FLOATING ASTRONAUT

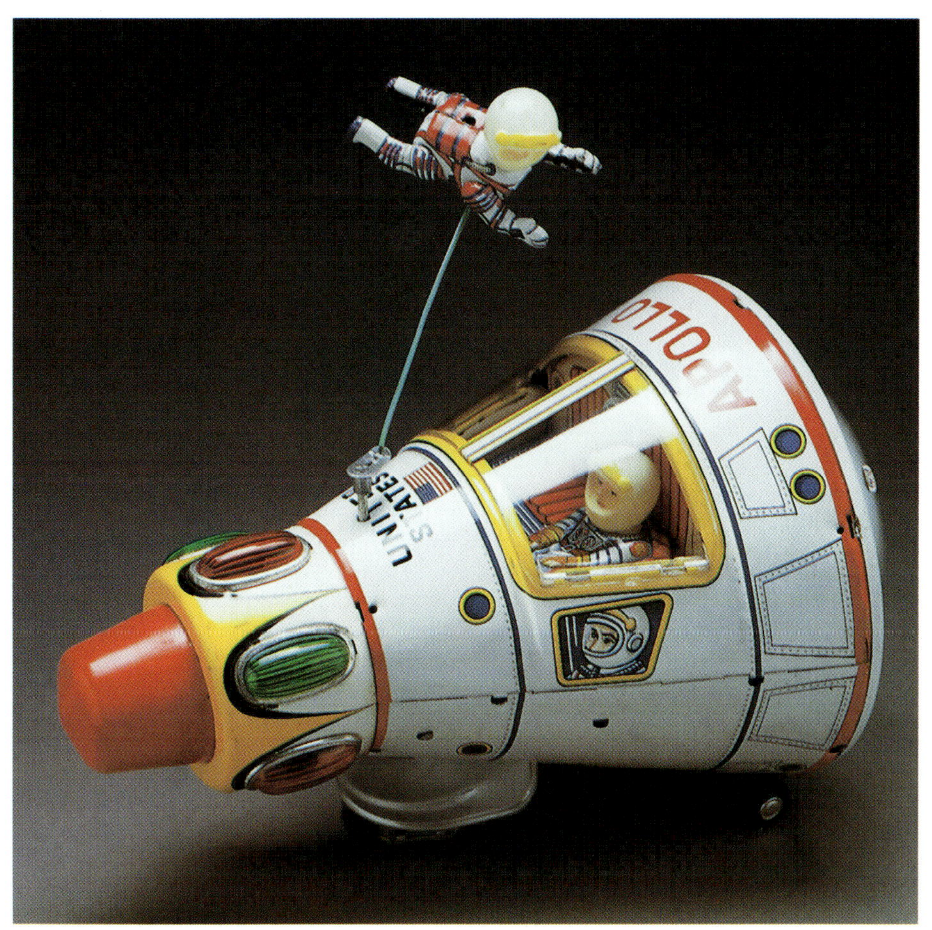

▲ 1960s, *Apollo Spacecraft*, Masudaya, 265 x 170 x 240 mm

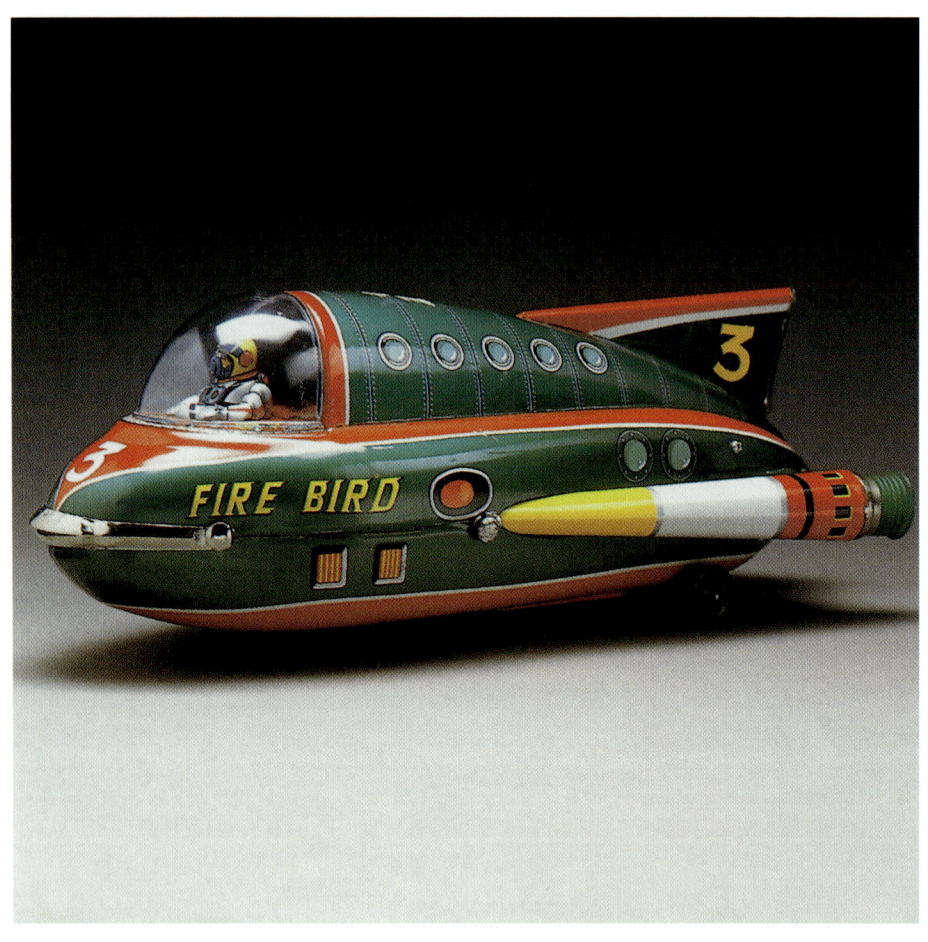

1960s, *Fire Bird*, Masudaya, 340 x 170 x 130 mm

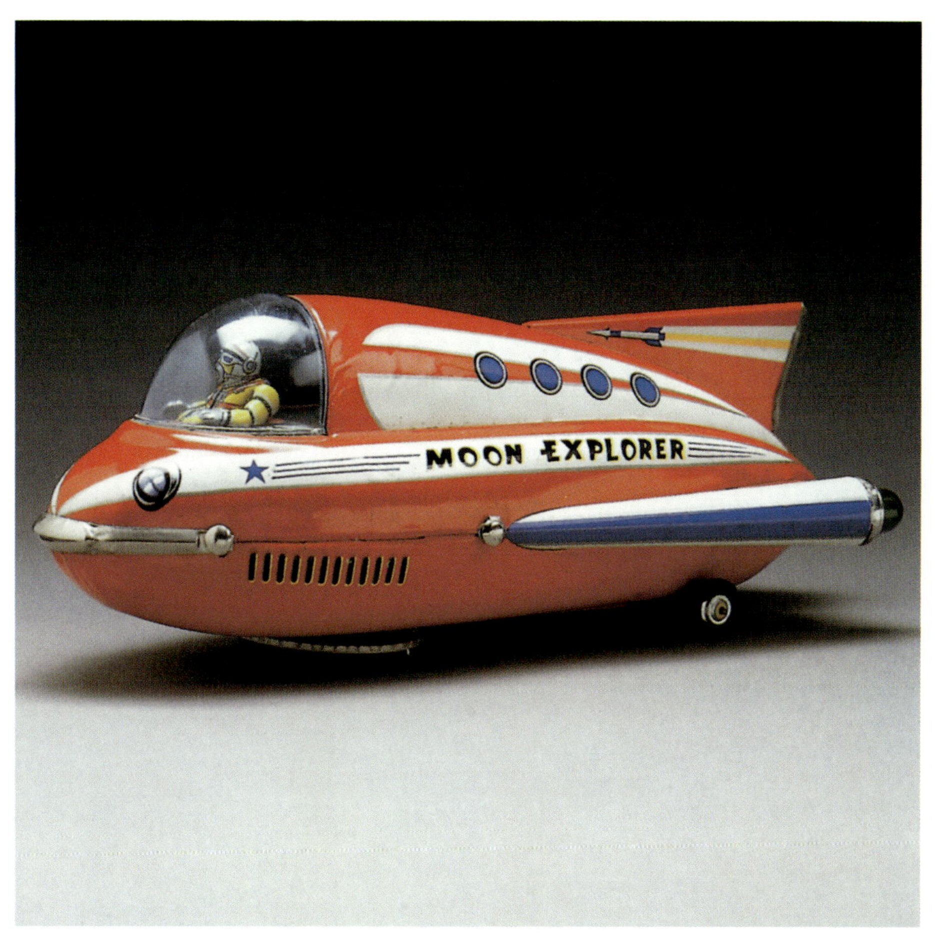

1960s, *Moon Explorer*, Masudaya, 340 x 170 x 130 mm

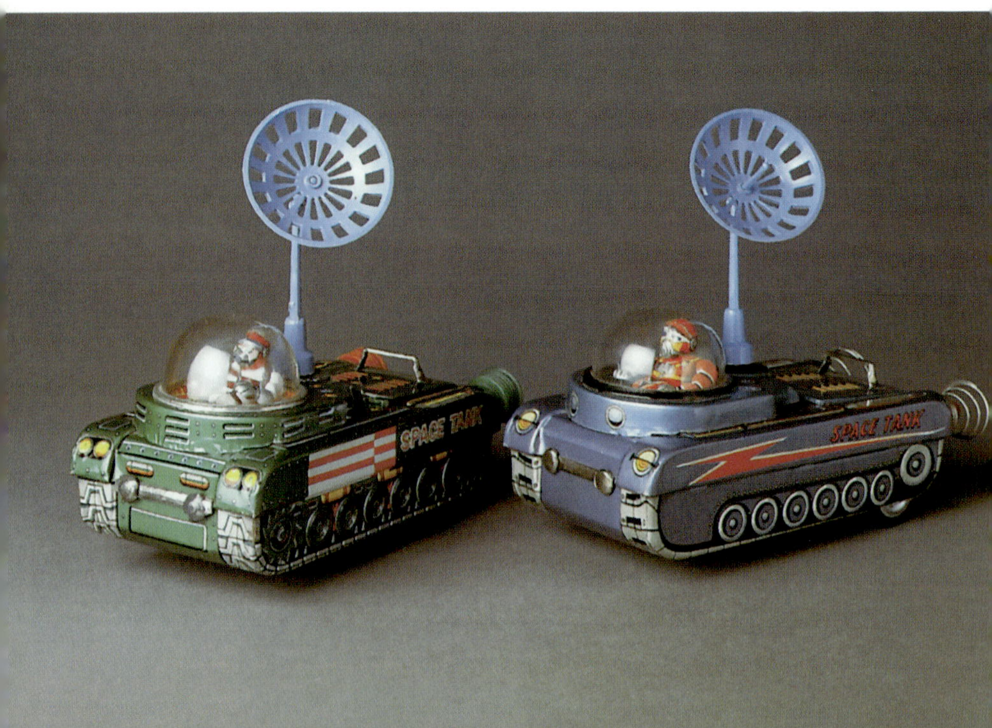

1960s, *Space Tank*, Masudaya, 215 x 105 x 200 mm

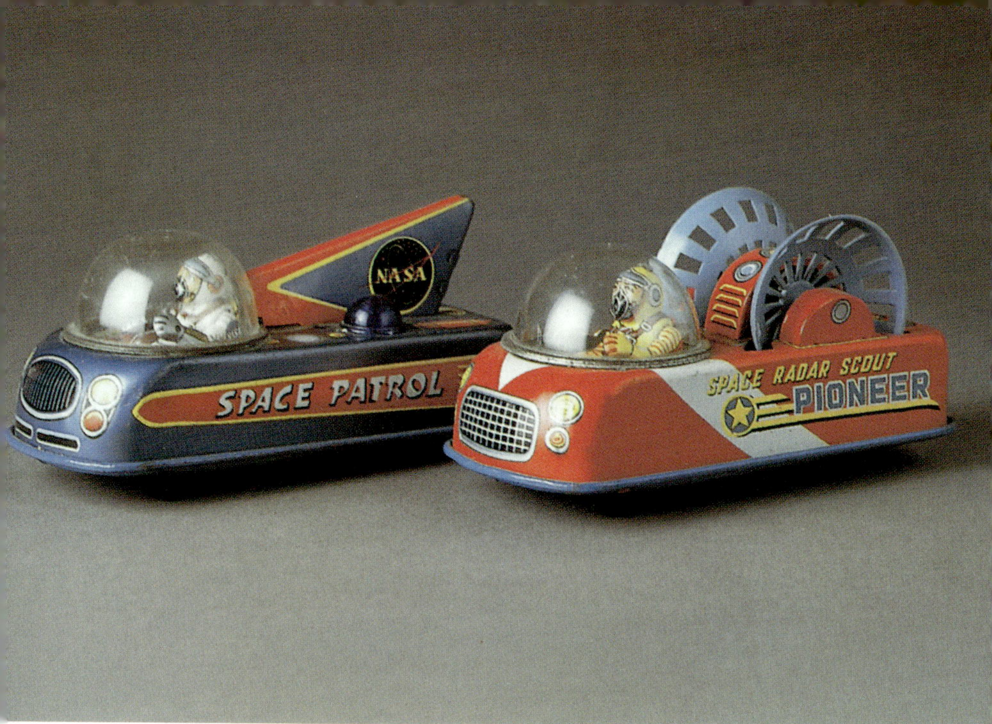

1960s, *Space Patrol*, unknown, 150 x 90 x 95 mm

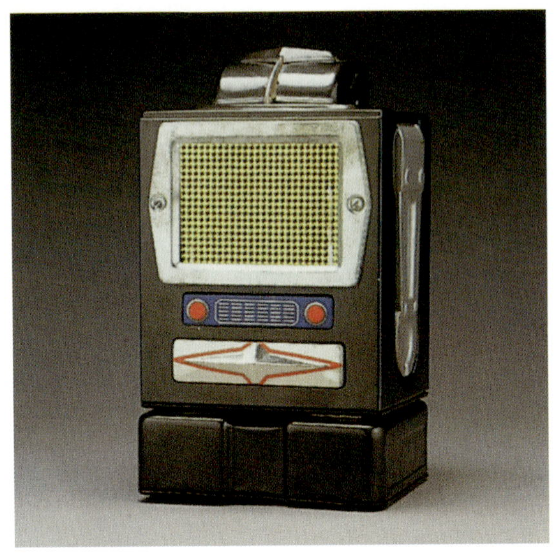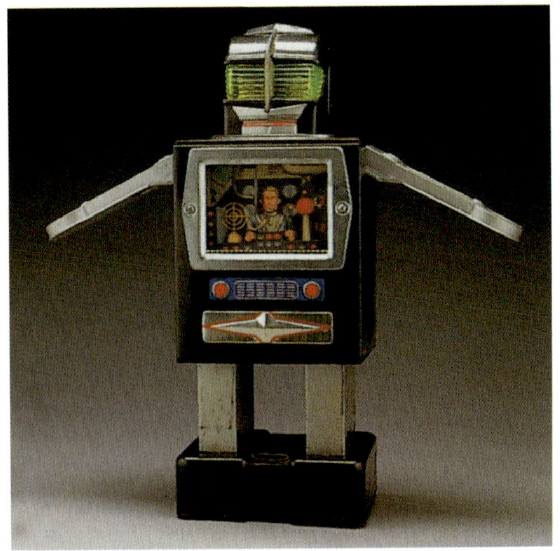

1950s, *Space Explorer*, Yonezawa, 92 x 110 x 290 mm

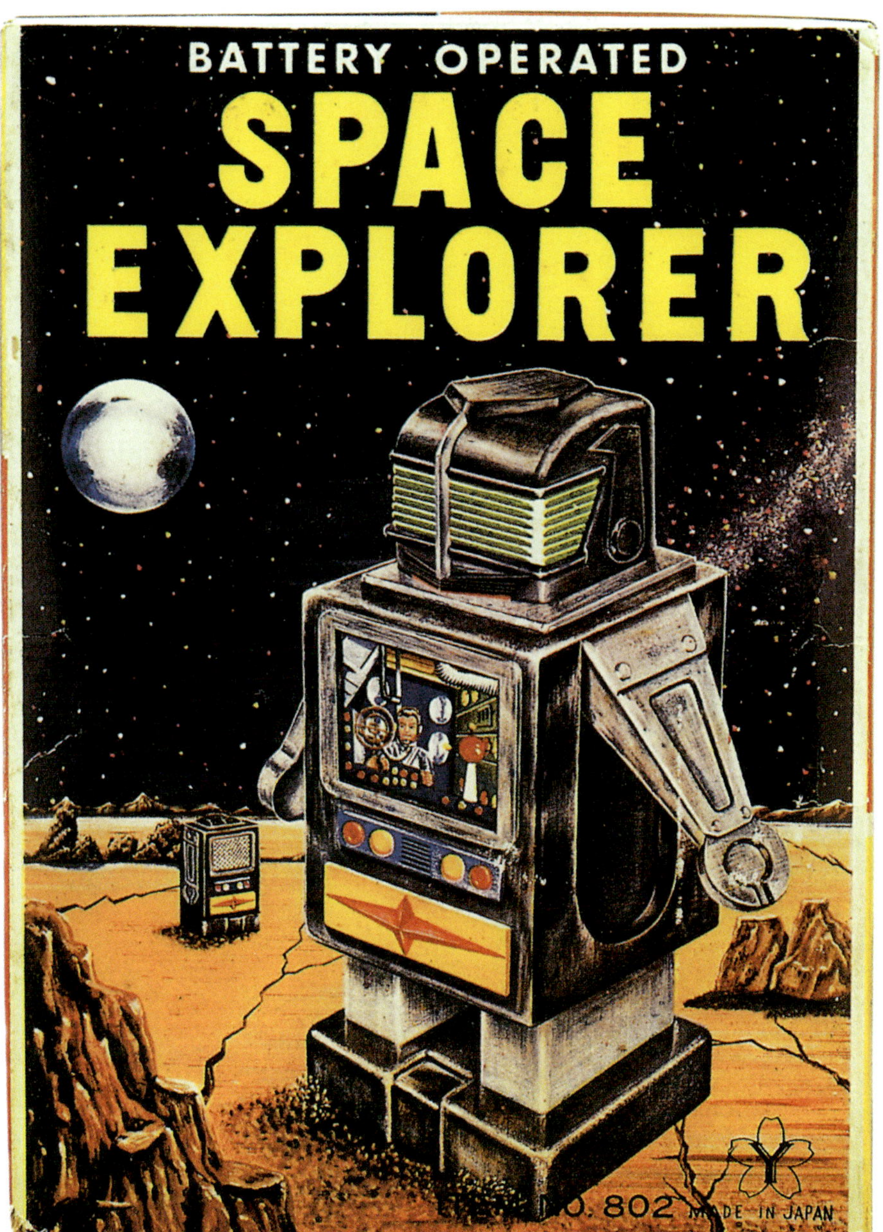
BATTERY OPERATED
SPACE
EXPLORER
NO. 802 MADE IN JAPAN

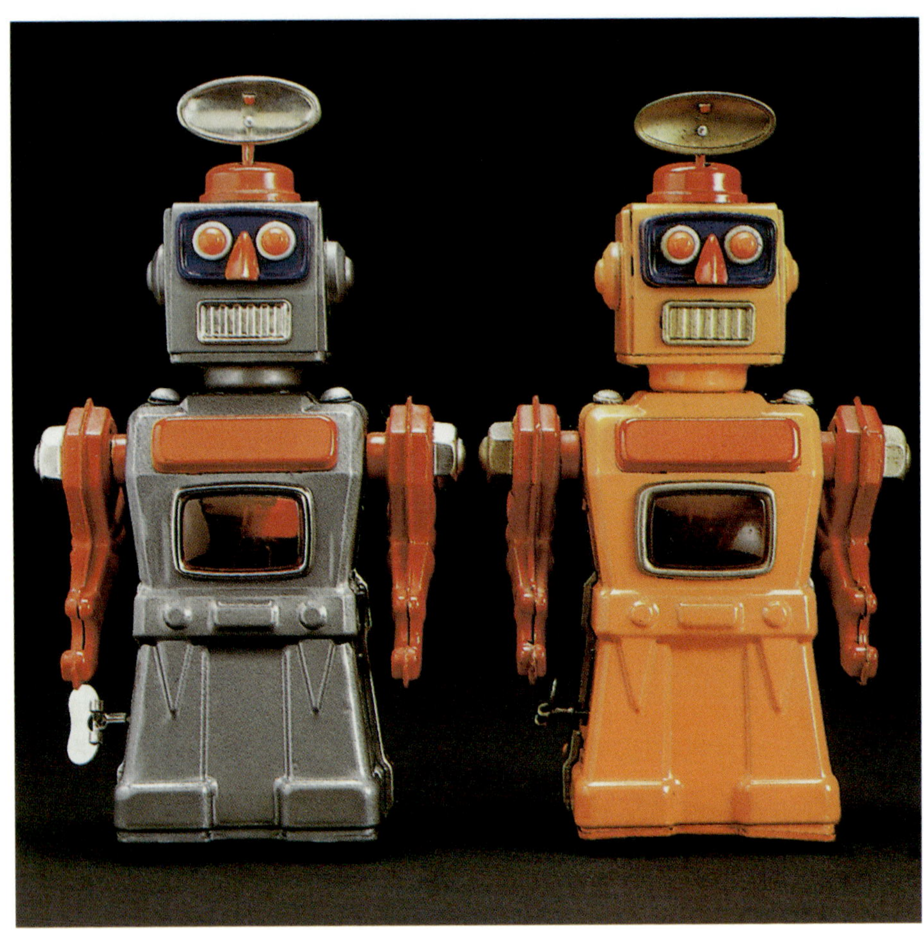

1950s, *Tremendous Mikes*, Aoshin, 90 x 155 x 260 mm

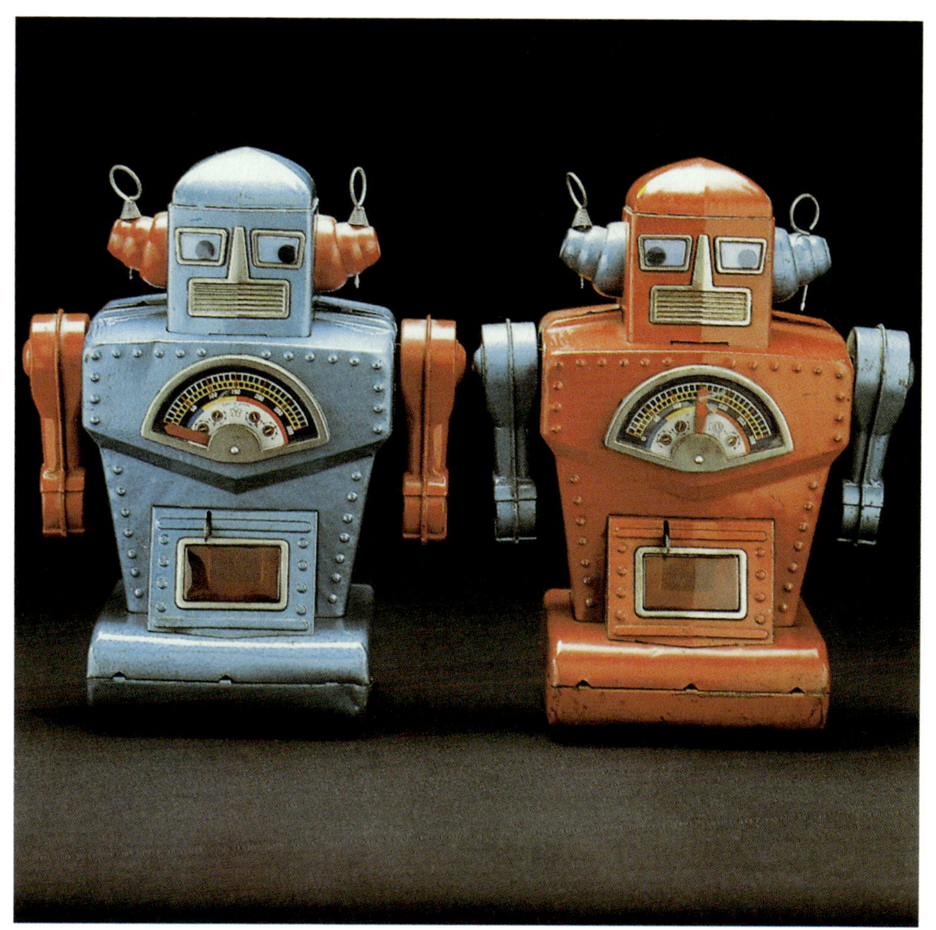

1950s, *Diamond Planet Robots*, Yonezawa, 140 x 200 x 260 mm

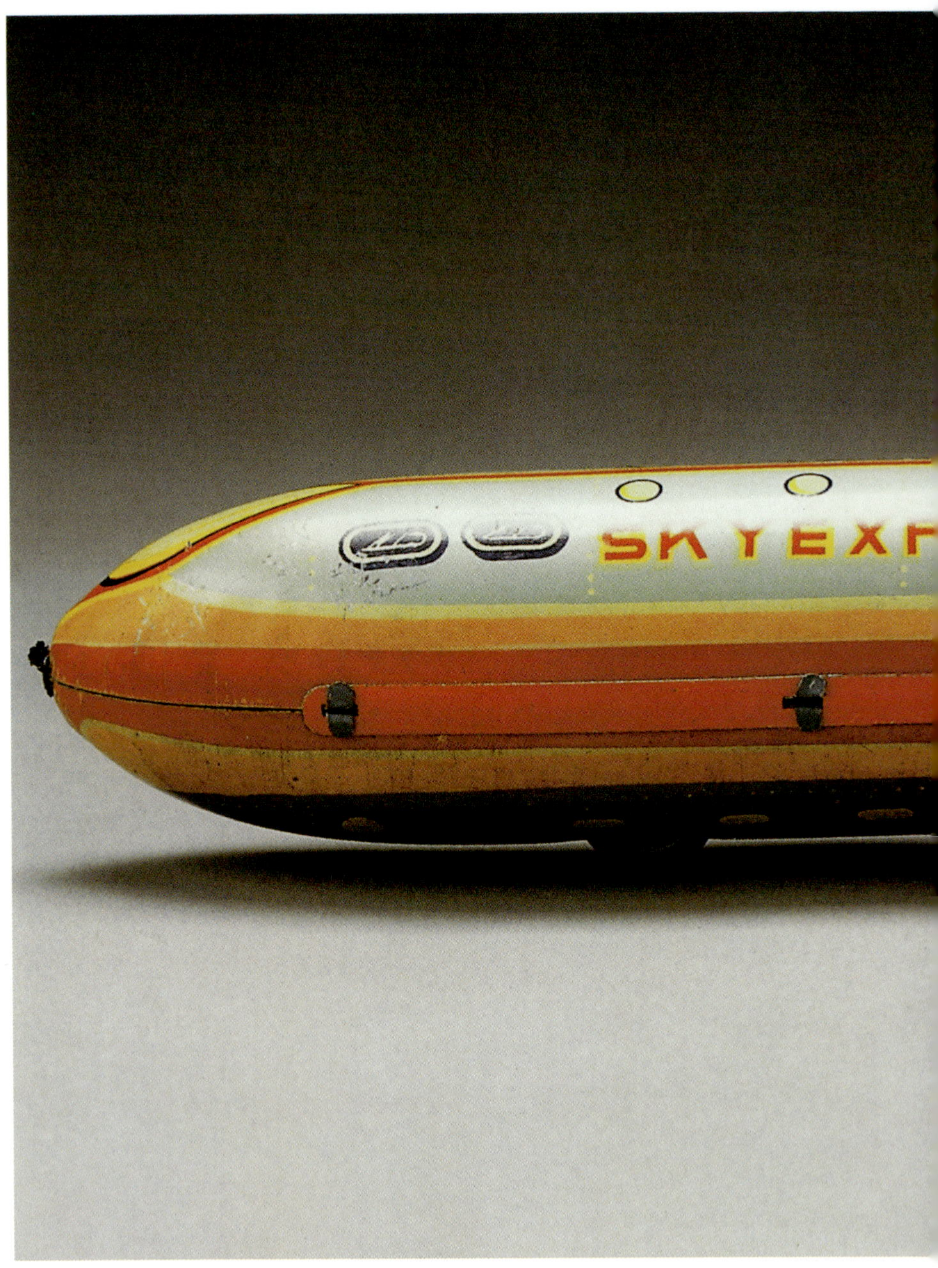

1950s, *Sky Express*, Usagiya, 218 x 58 x 60 mm

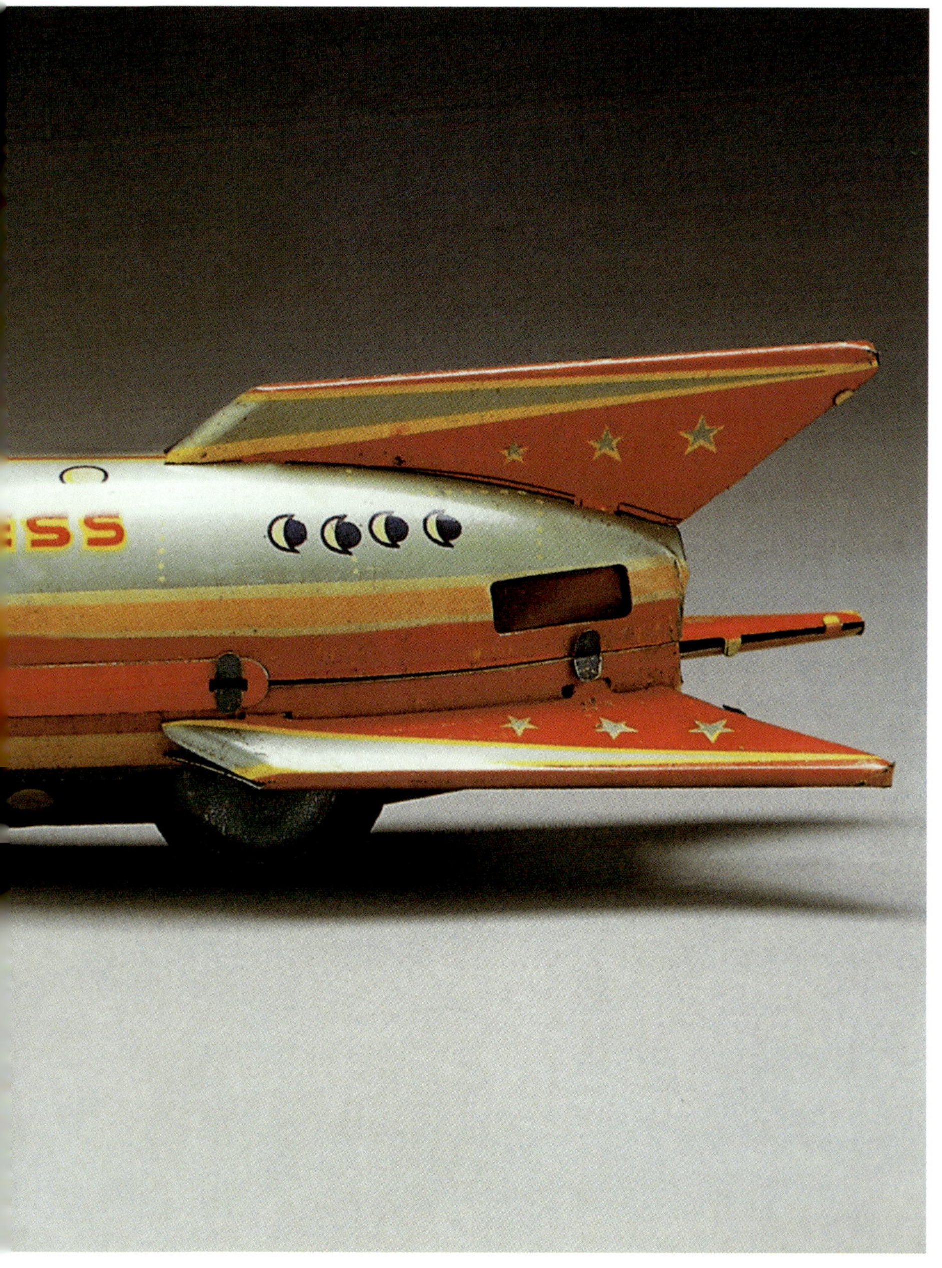

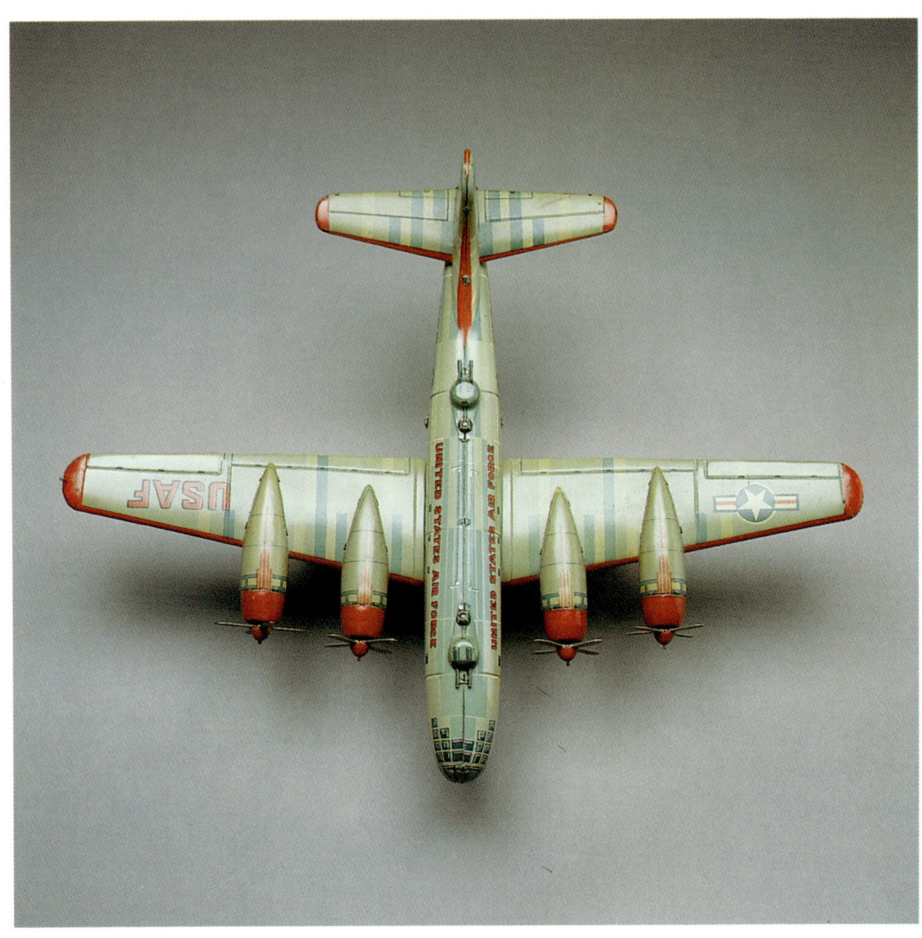

1960s, *Airplane*, Yonezawa, 360 x 285 x 110 mm

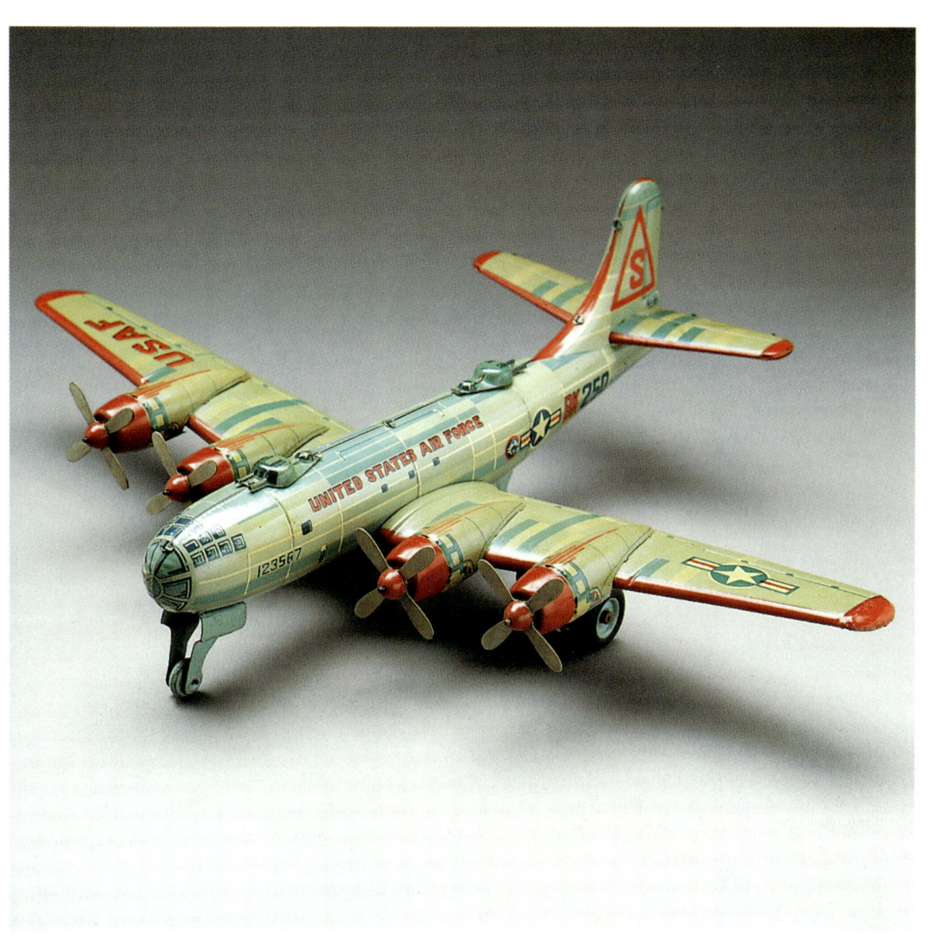

USAF
UNITED STATES AIR FORCE
123567
BK 250
S

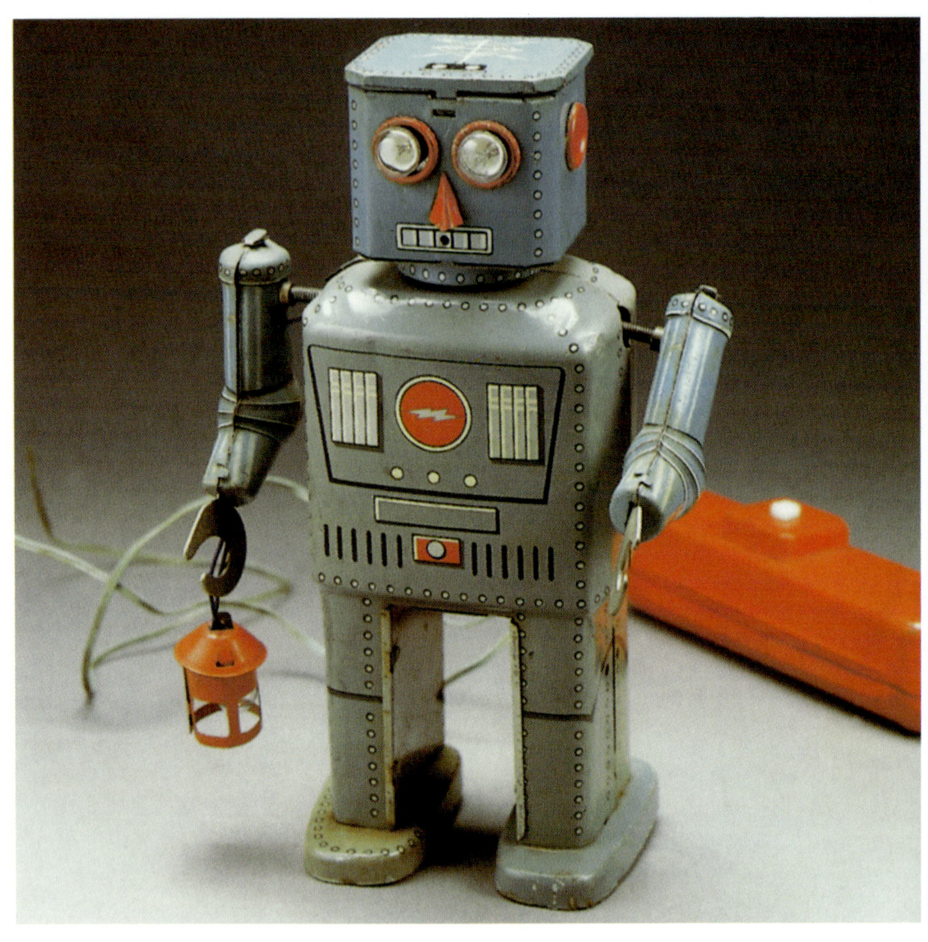

1950s, *Robot with Lantern*, Line Mar, 90 x 115 x 200 mm

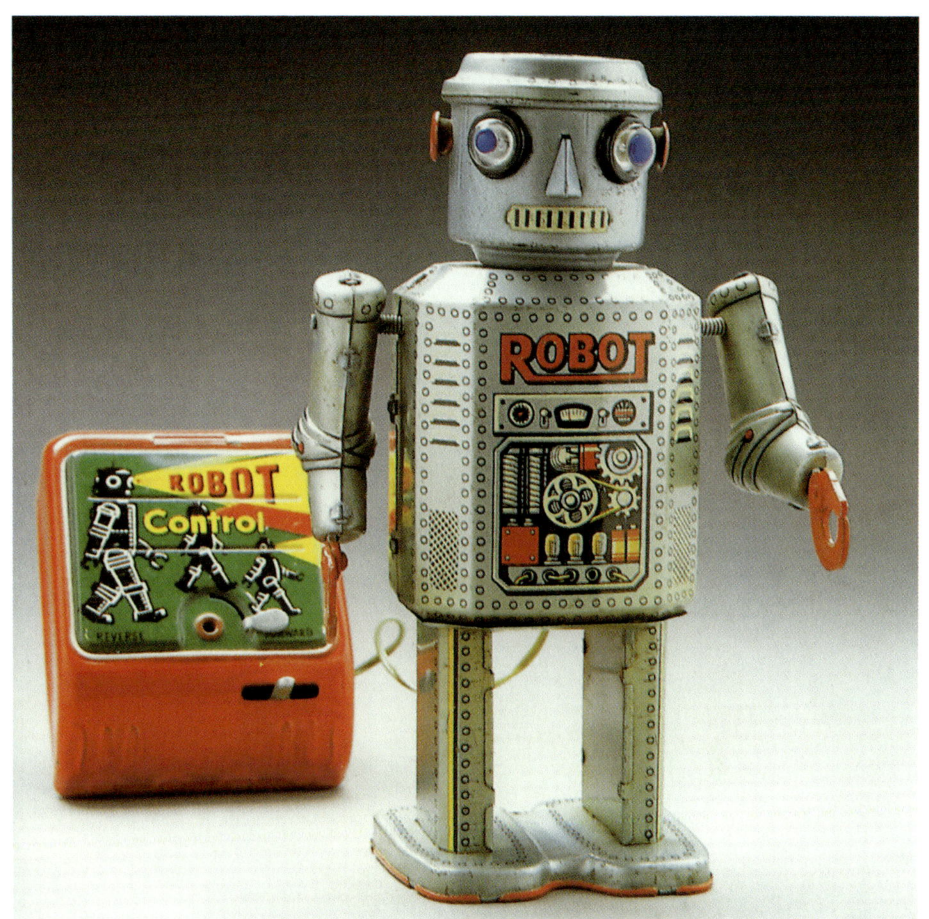

1950s, *Robot*, Masudaya, 65 x 110 x 190 mm

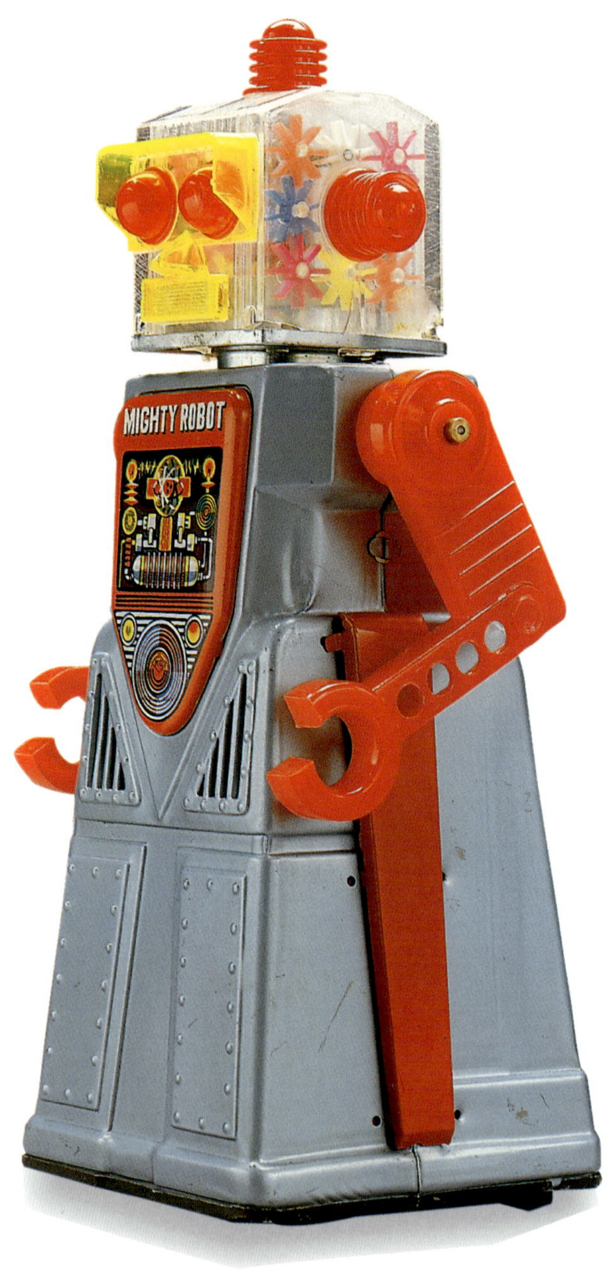

1960s, *Mighty Robot*, Yoshiya, 110 x 150 x 300 mm

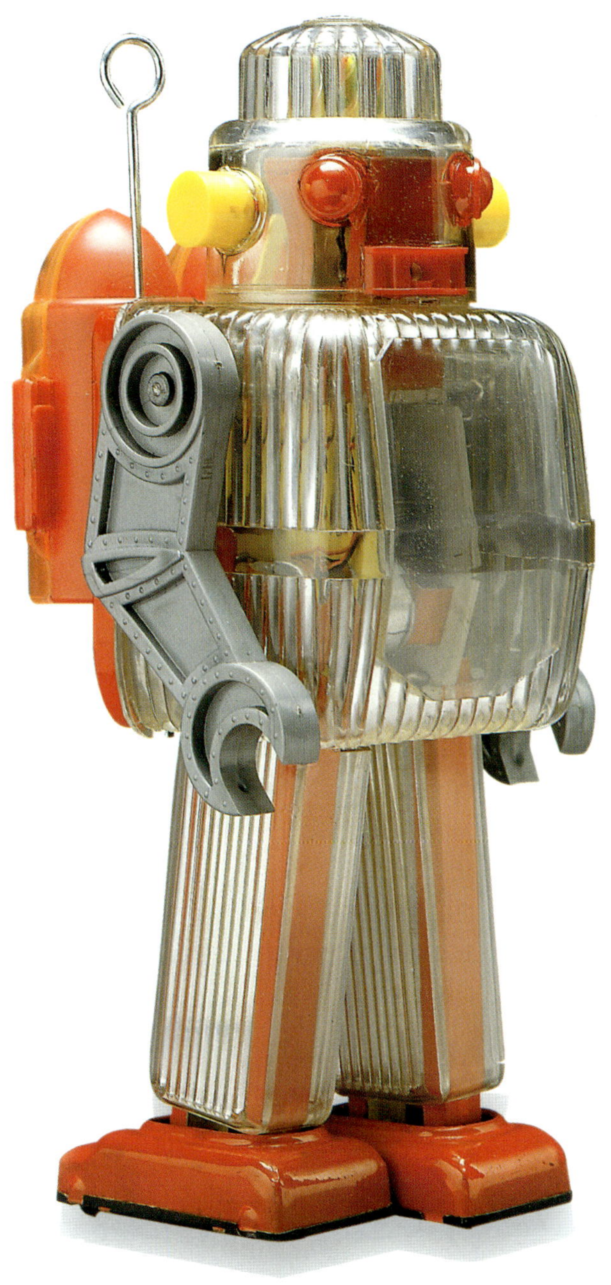

1950s, *Ranger Robot*, Daiya, 115 x 110 x 265 mm

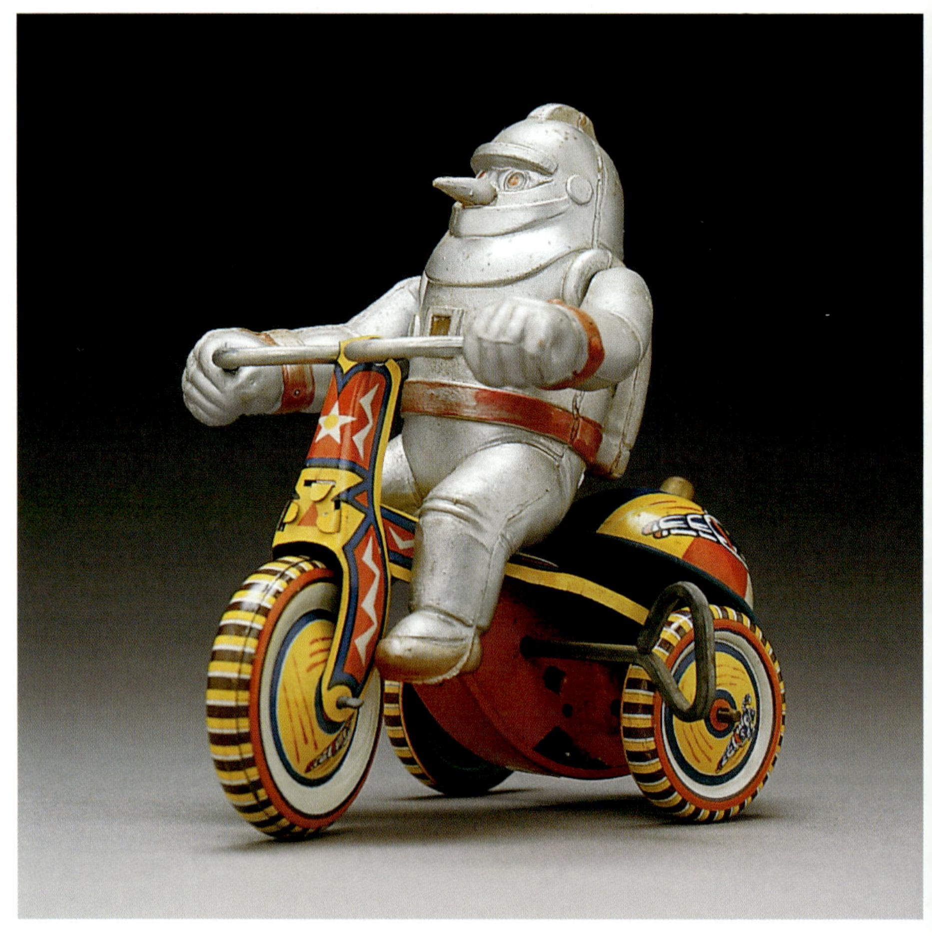

▲ 1950s, *Tetsujin 28-GO*, Miura, 112 x 78 x 118 mm

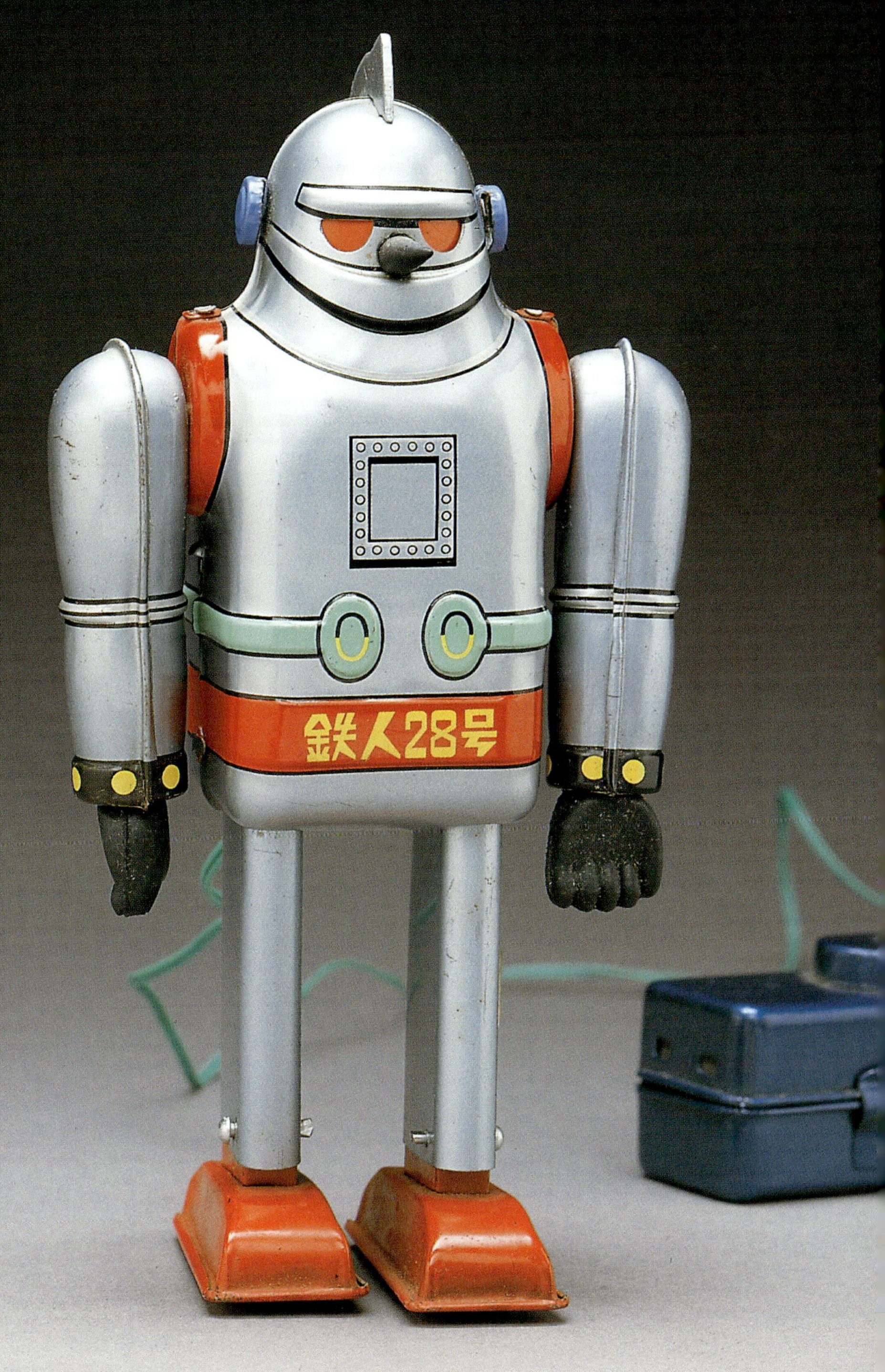

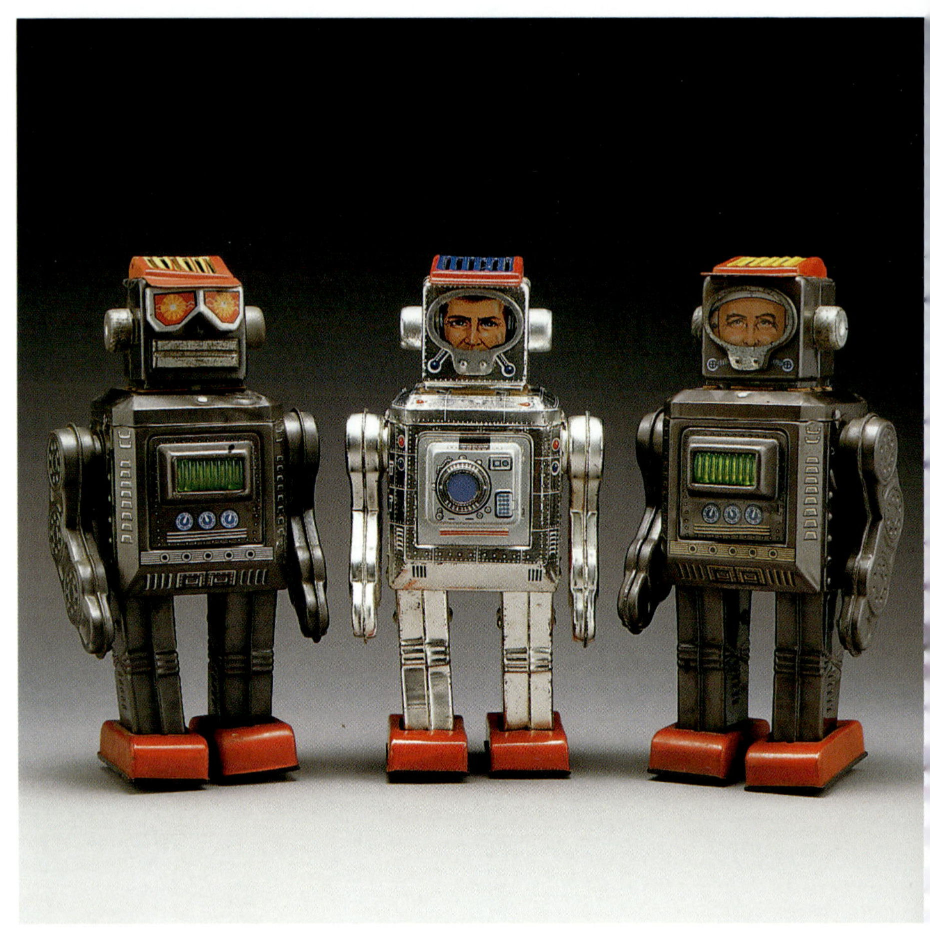

1960s, *Robots*, unknown, 82 x 105 x 220 mm

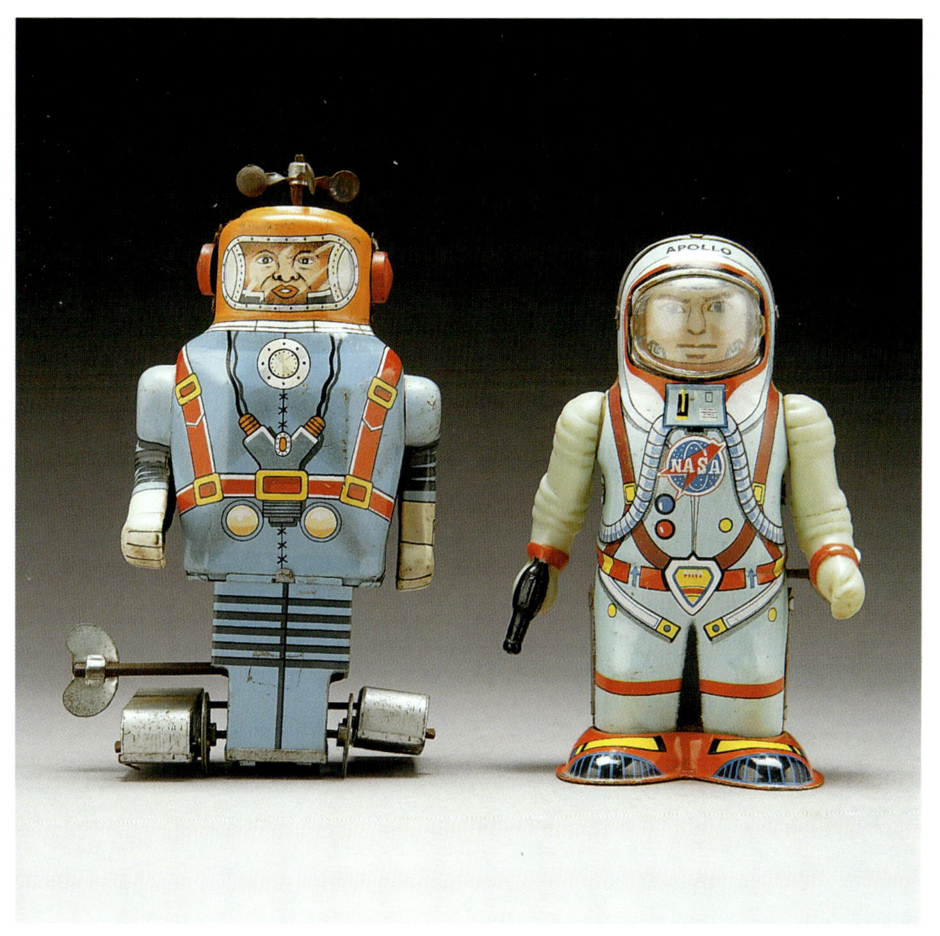

1950s, 1960s, *Astronauts,* SY Toys, Shudo, 50 x 90 x 145, 45 x 85 x 128 mm

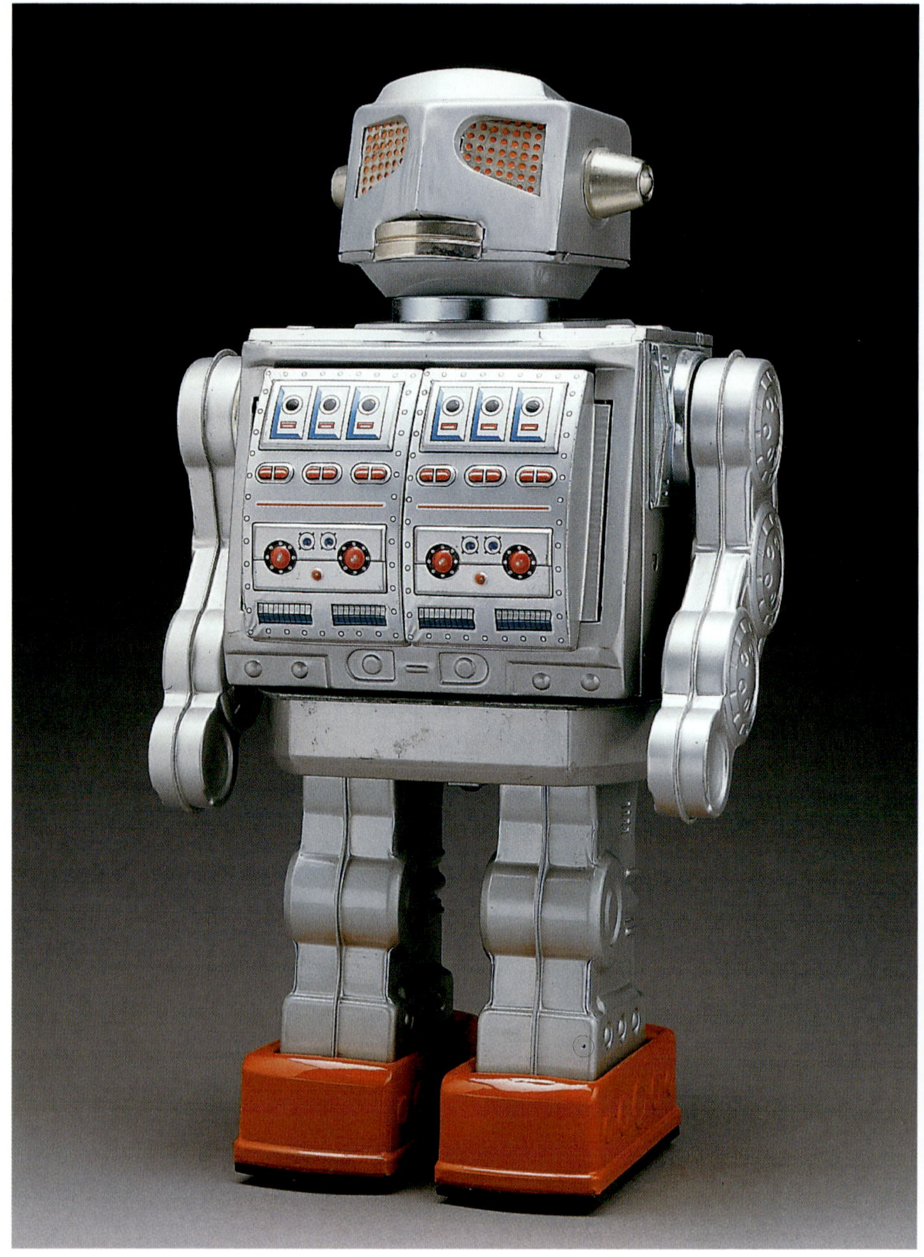

1950s, *Giant Robot*, Horikawa, 148 x 220 x 400 mm

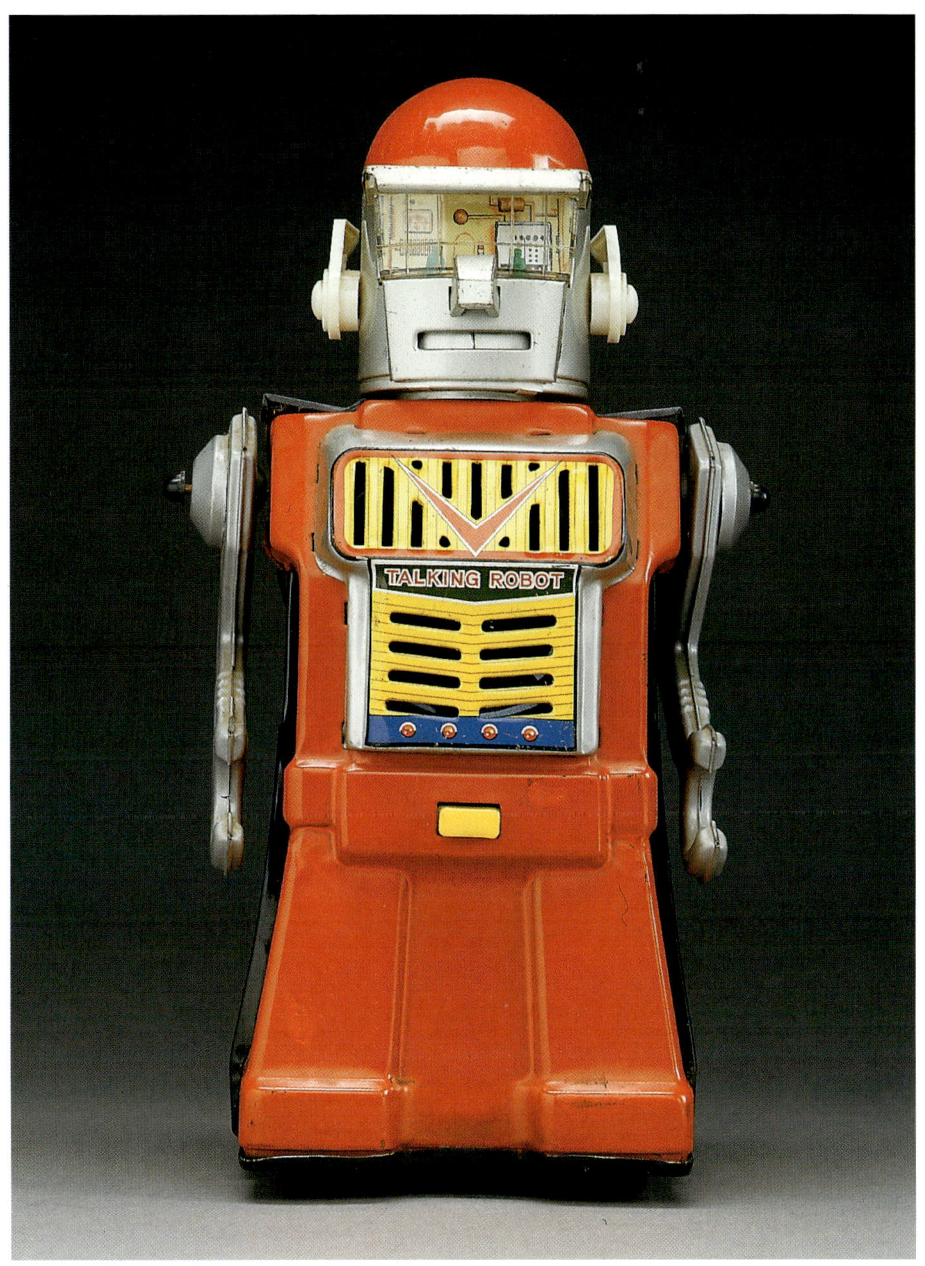

1960s, *Talking Robot*, Yonezawa, 112 x 150 x 275 mm

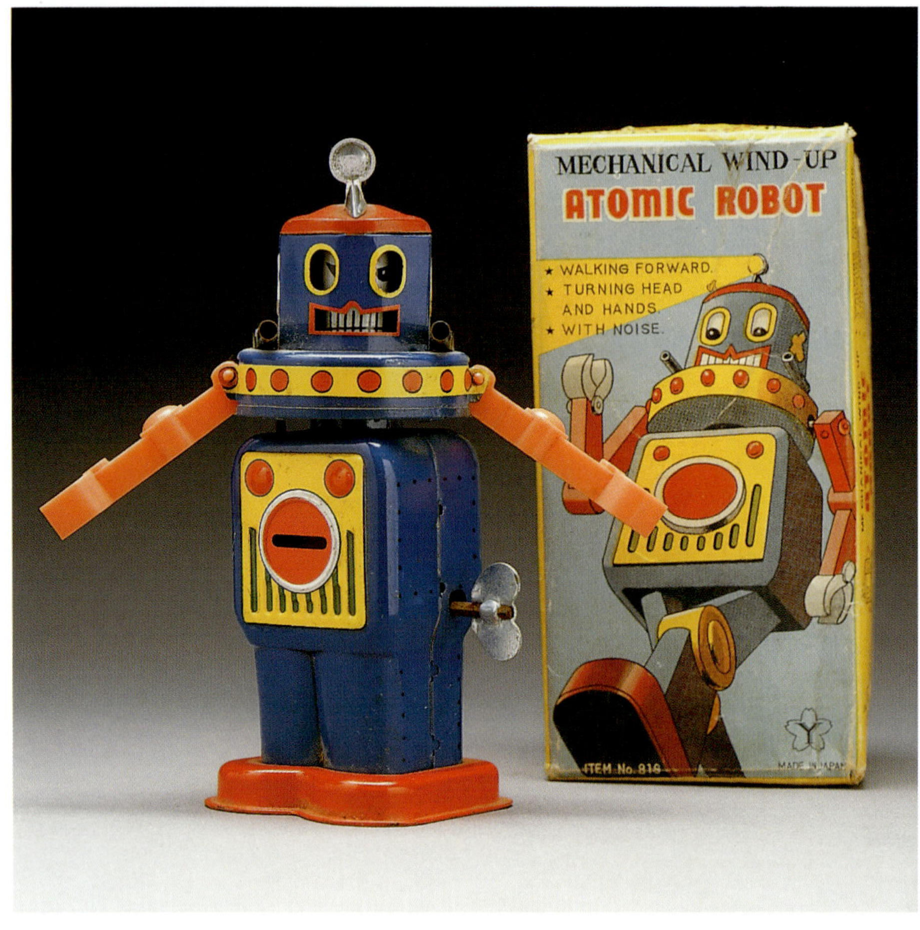

1960s, *Atomic Robot and Case*, Yonezawa, 55 x 140 x 160 mm

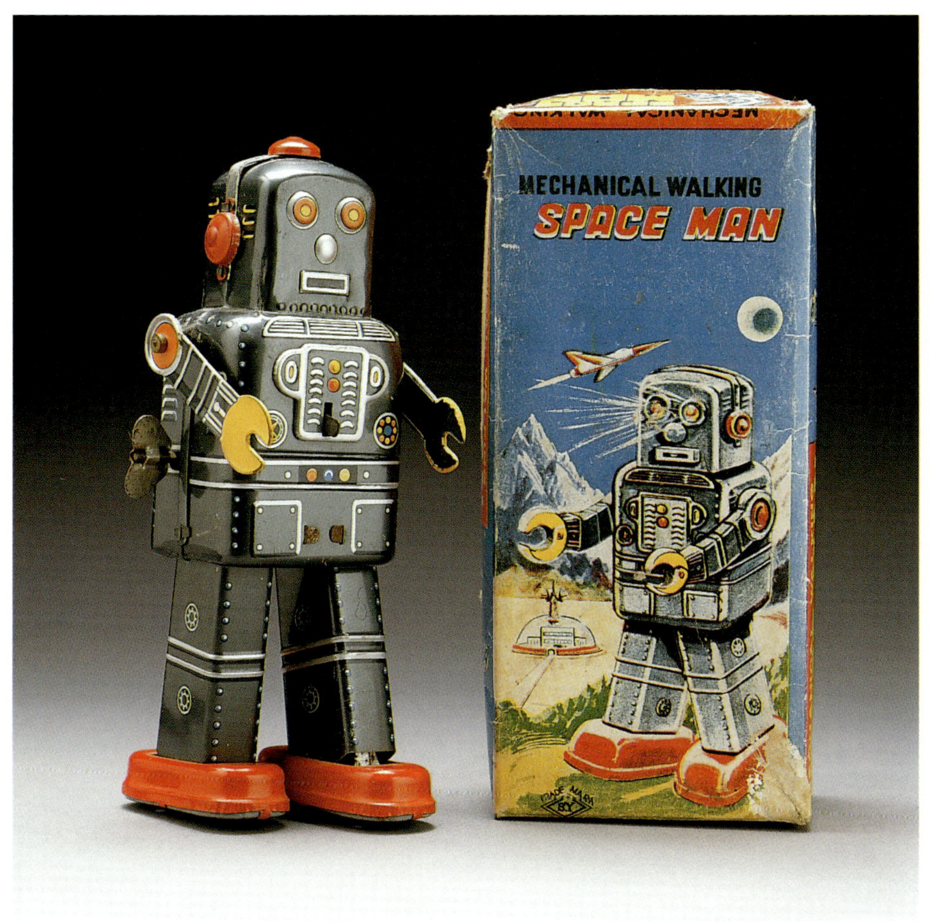

1950s, *Space Man and Case*, SY Toys, 64 x 80 x 195 mm

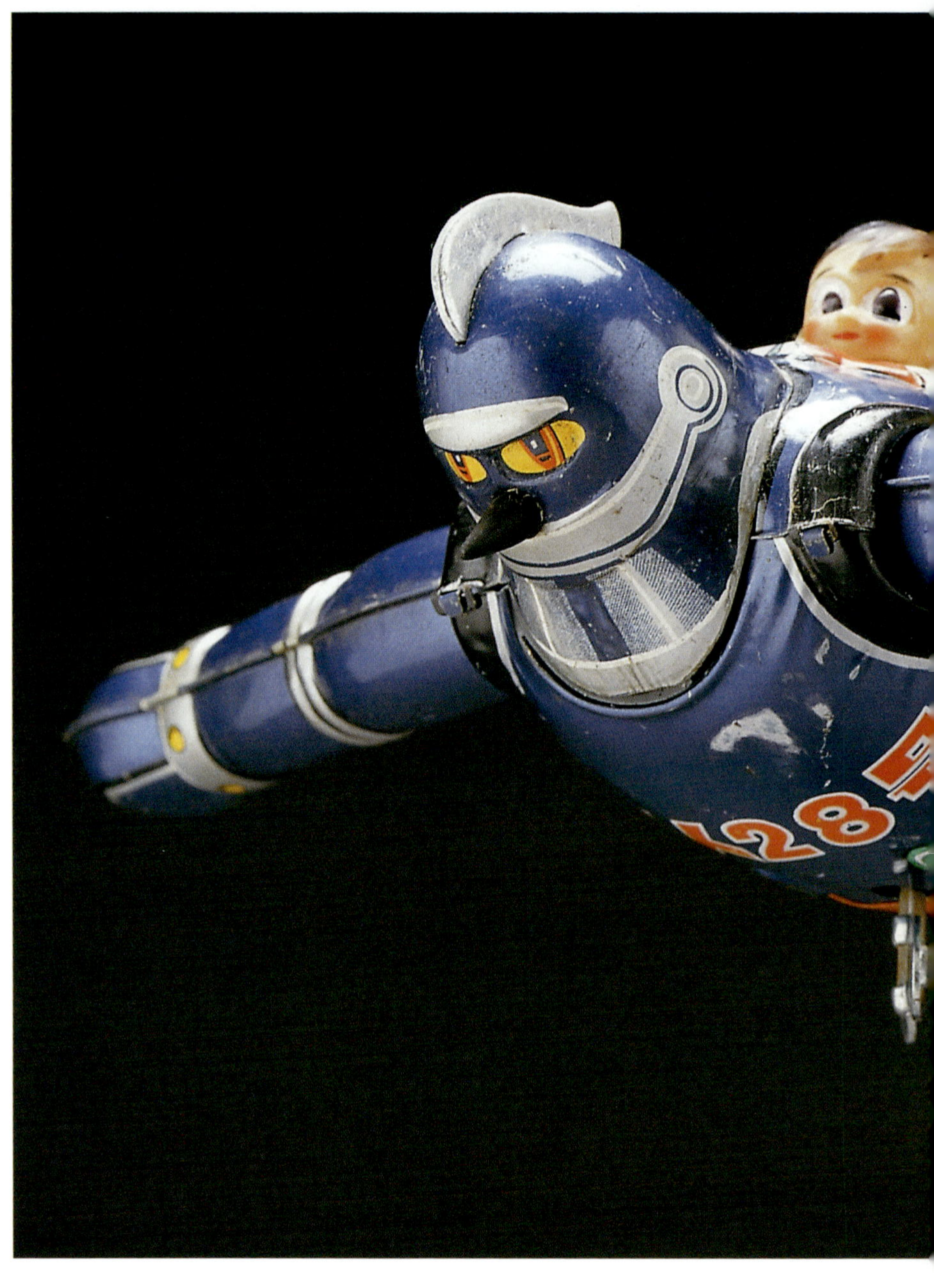

1960s, *Tetsujin 28-GO*, Bandai, 220 x 115 x 168 mm

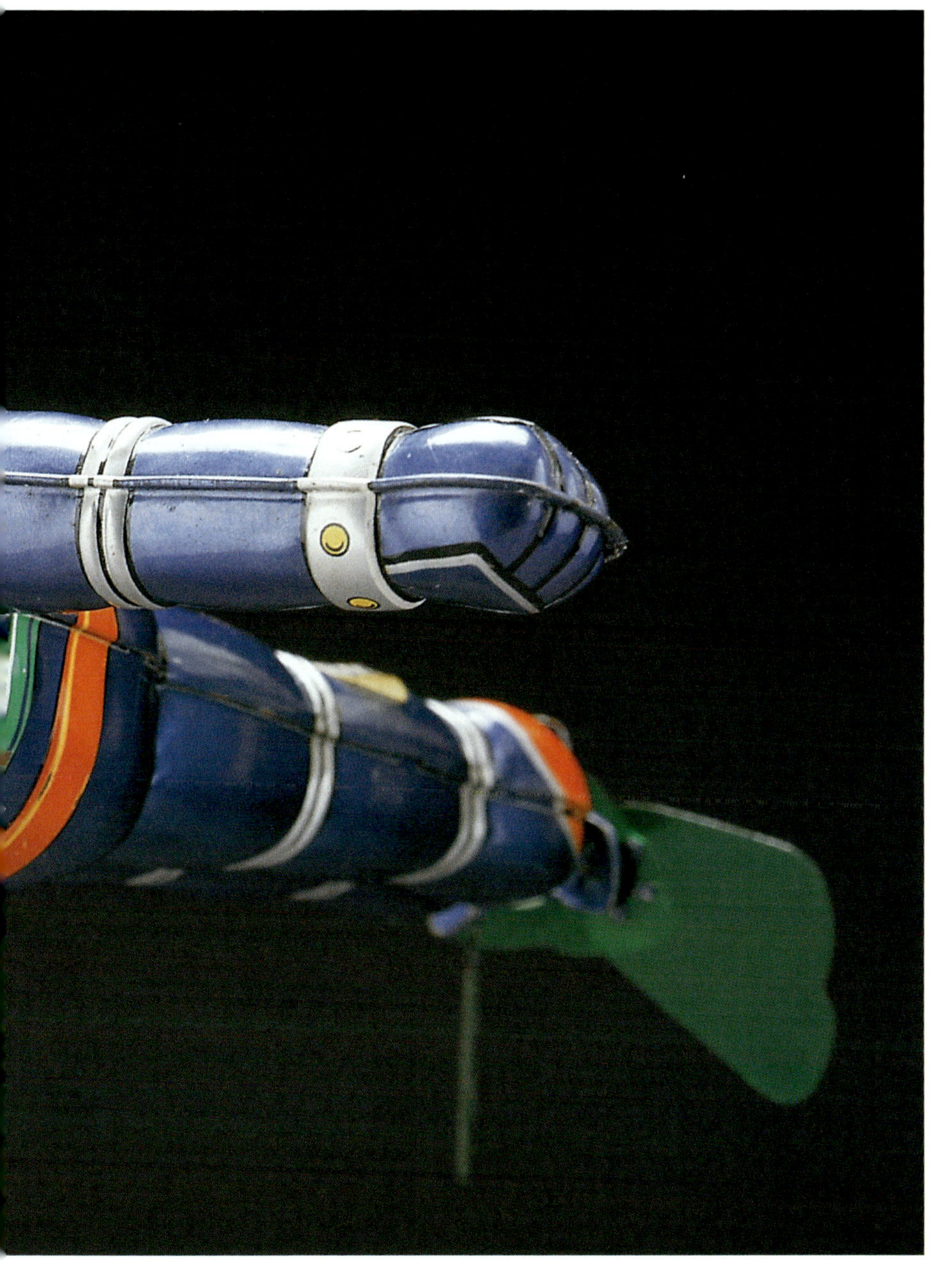

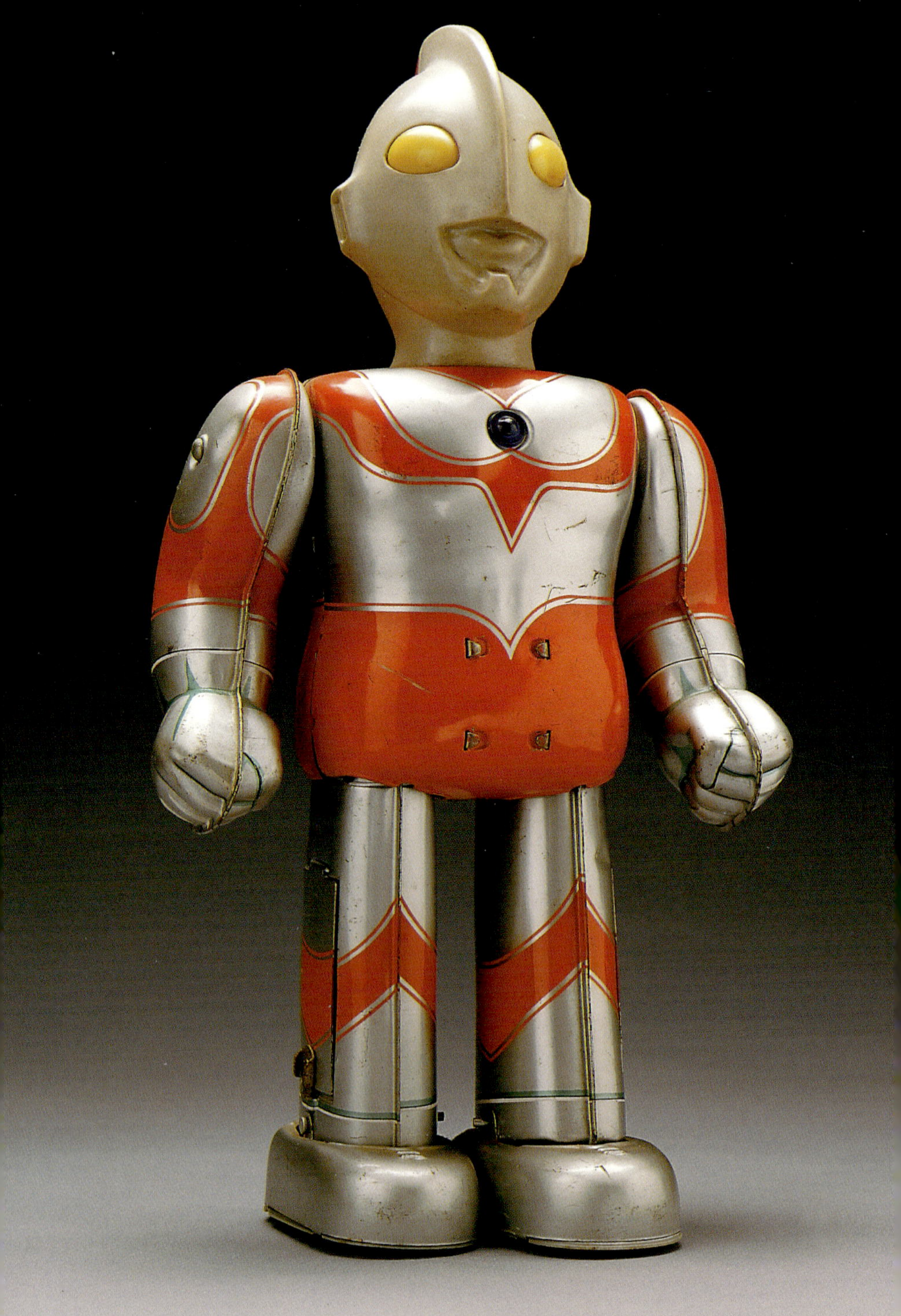

◄1960s, *Ultra Man*, Bull Mark, 90 x 150 x 320 mm

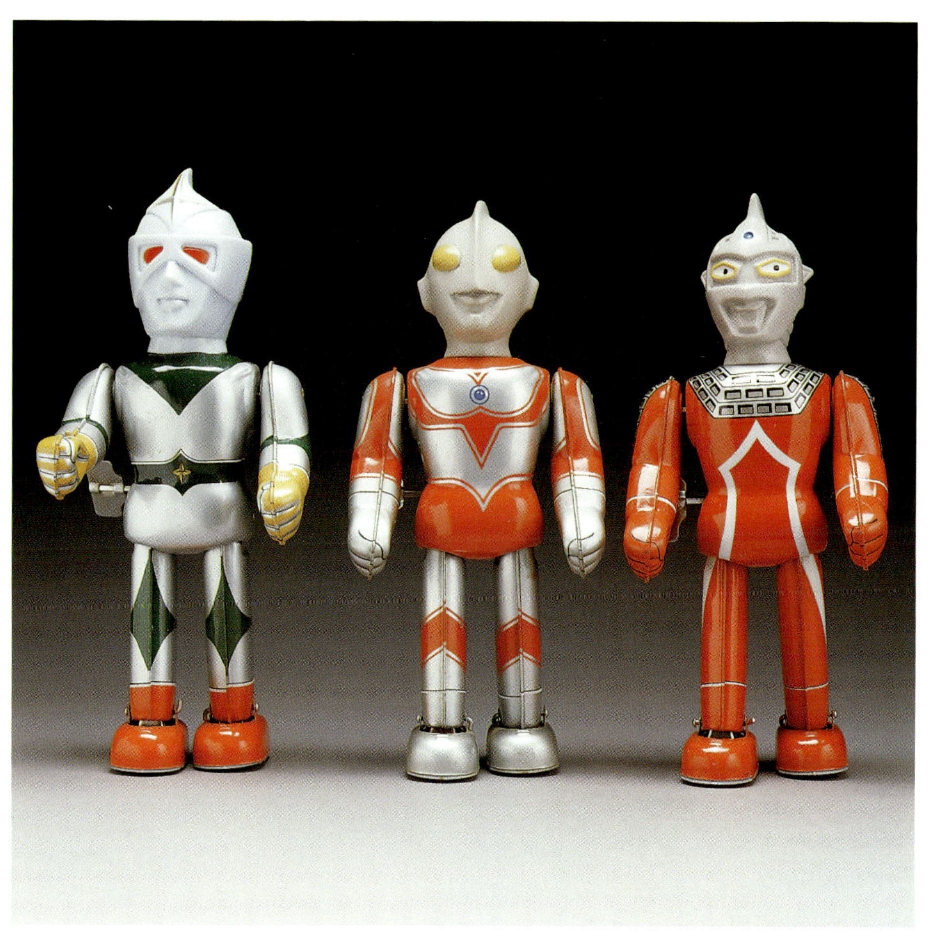

▲1960s, *Mirror Man, Ultra Man, Ultra 7*, Bull Mark, 50 x 100 x 235 mm

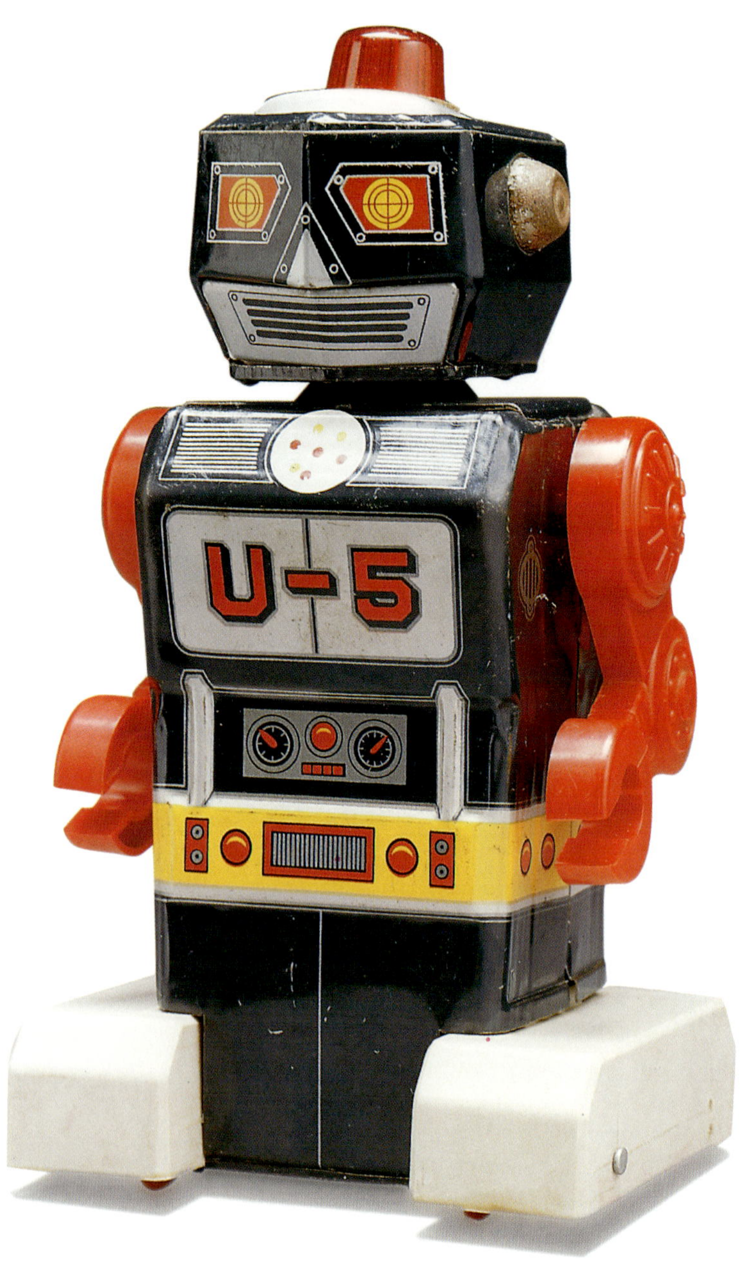

1960s, *U-5 Robot*, Daiya, 83 x 105 x 193 mm

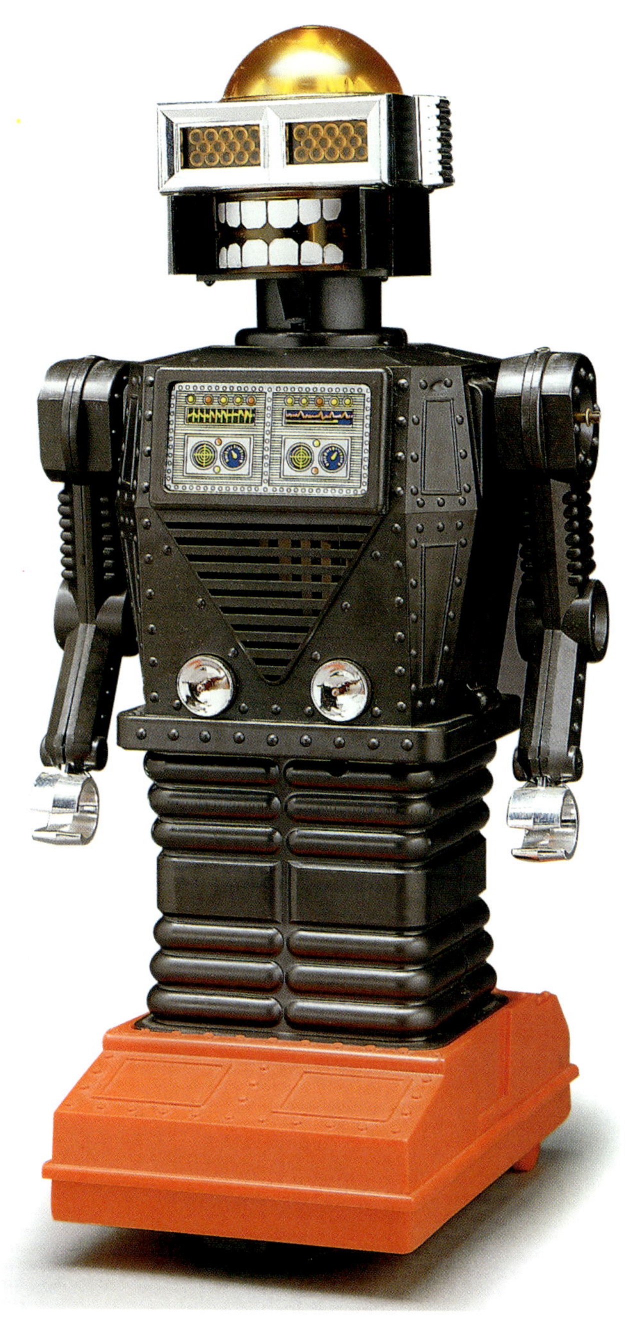

1960s, *Hysterical Robot*, unknown, 165 x 160 x 340 mm

1960s, *Robotank R-1*, Nomura, 195 x 140 x 260 mm

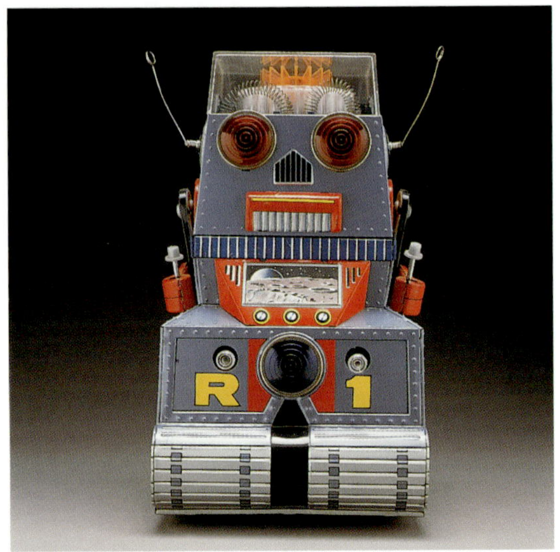

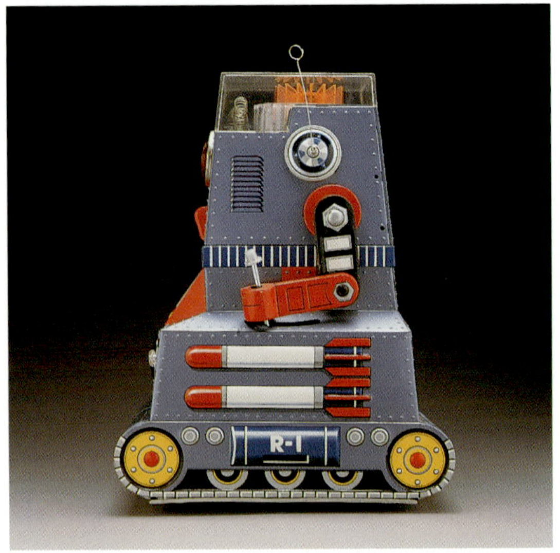

1960s, *Robotank-Z*, Nomura, 195 x 140 x 260 mm

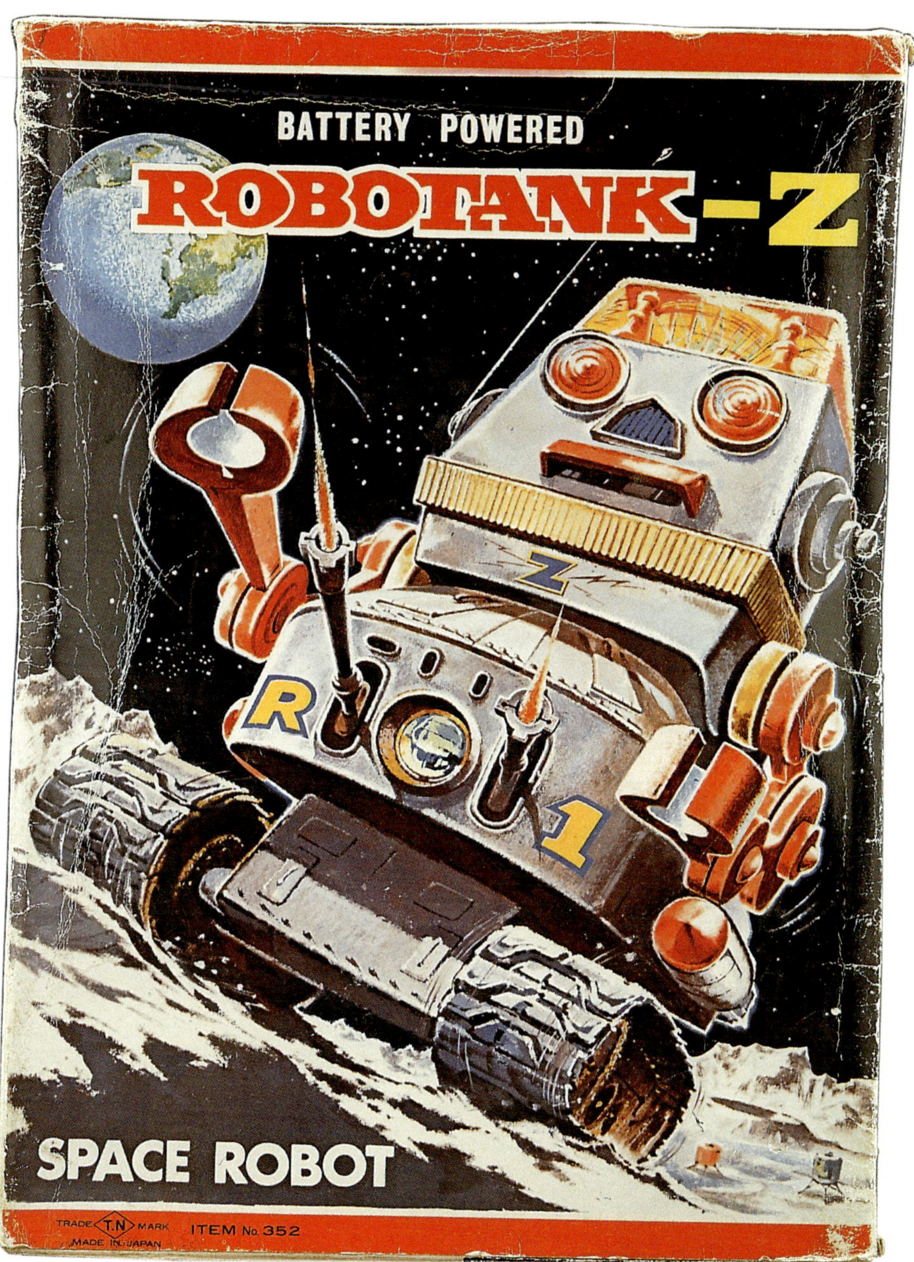
BATTERY POWERED
ROBOTANK-Z
R
Z
1
SPACE ROBOT
TRADE T.N MARK
MADE IN JAPAN
ITEM No. 352

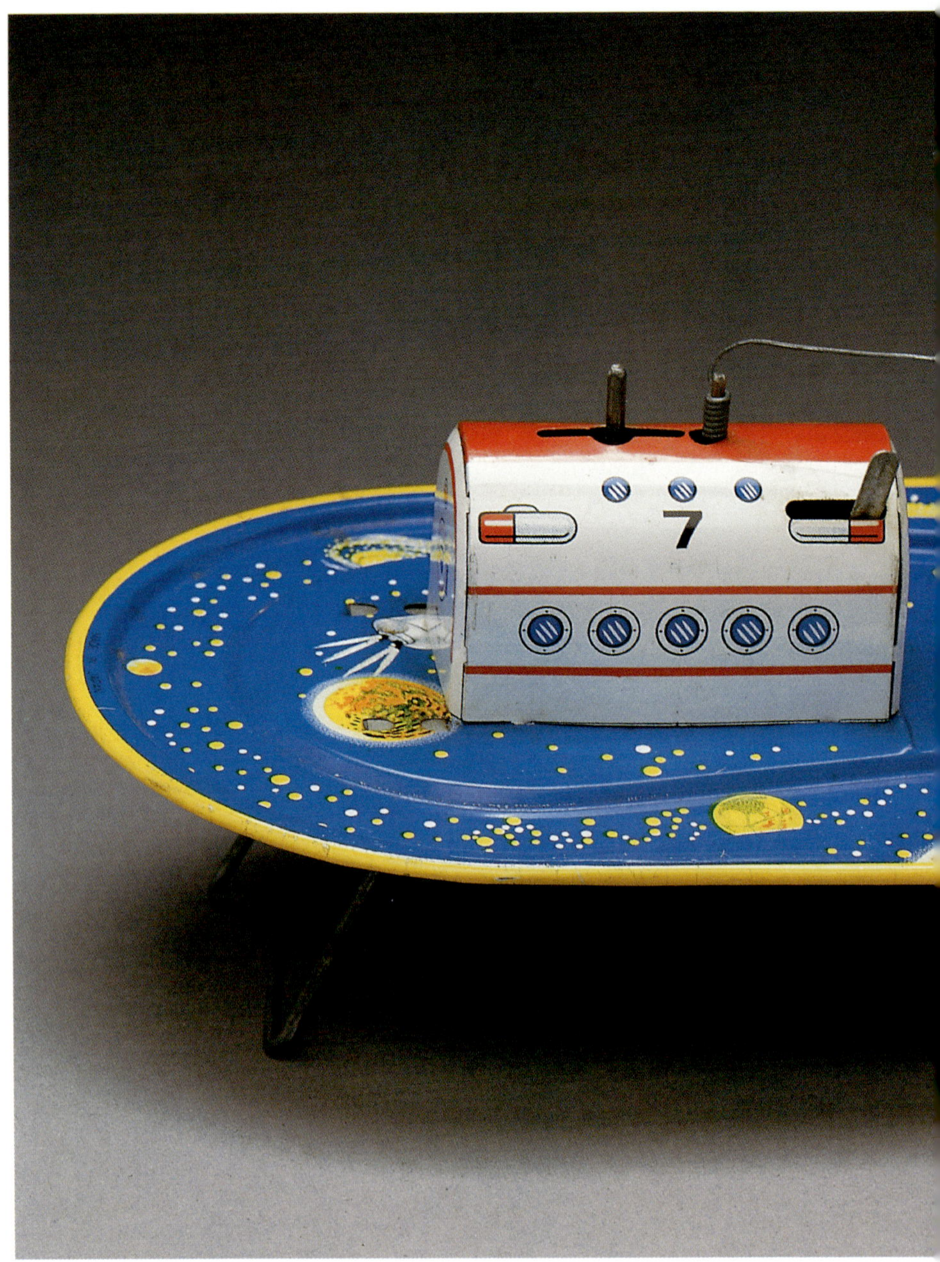

1960s, *Tetsujin 28-GO*, unknown, 273 x 138 x 150 mm

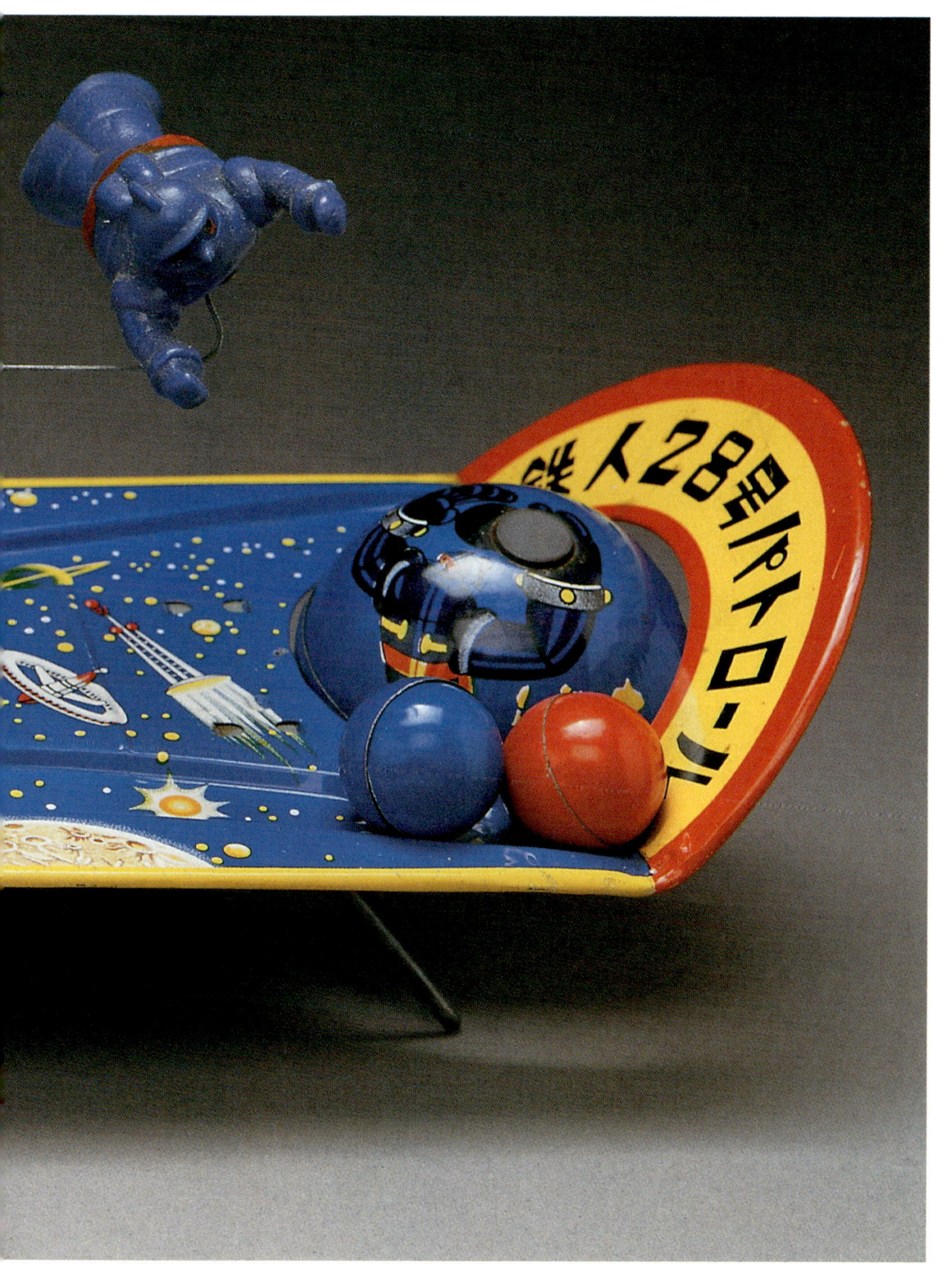

1960s, *Mike Robot*, Tomy, 110 x 90 x 305 mm ▶

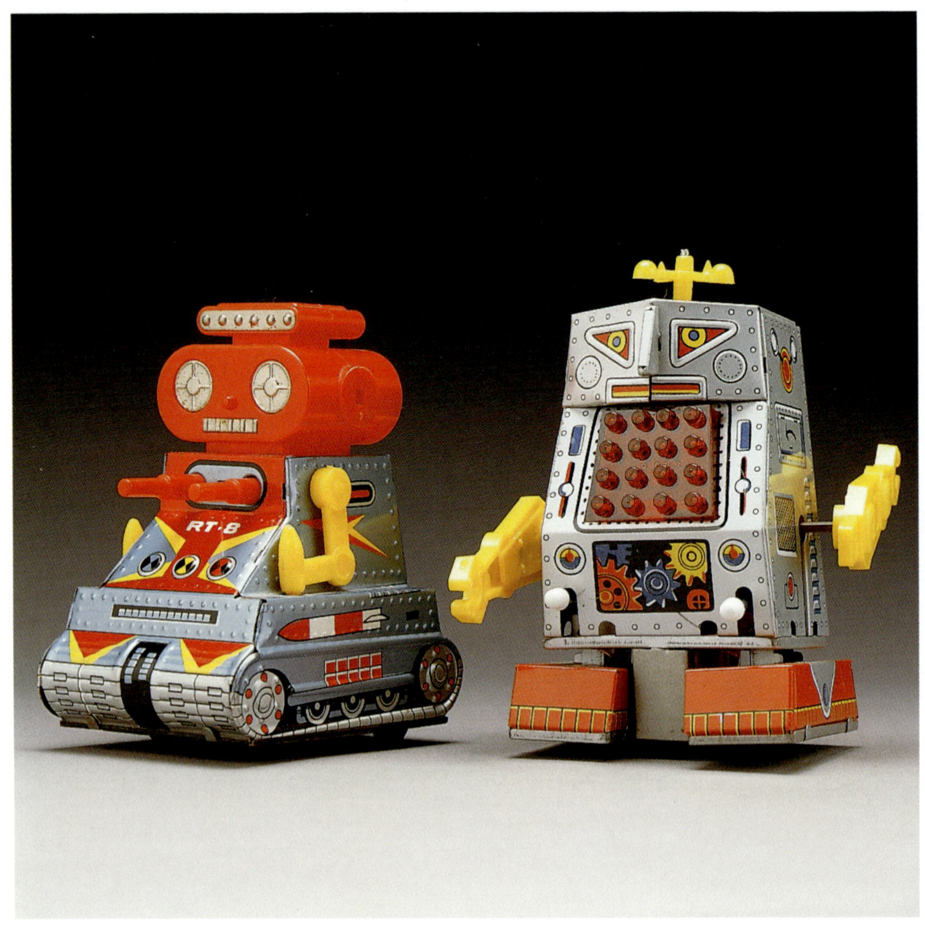

▲ 1960s, *RT-8 Robot, Robot*, Nomura, unknown, 95 x 73 x 120, 80 x 95 x 135 mm

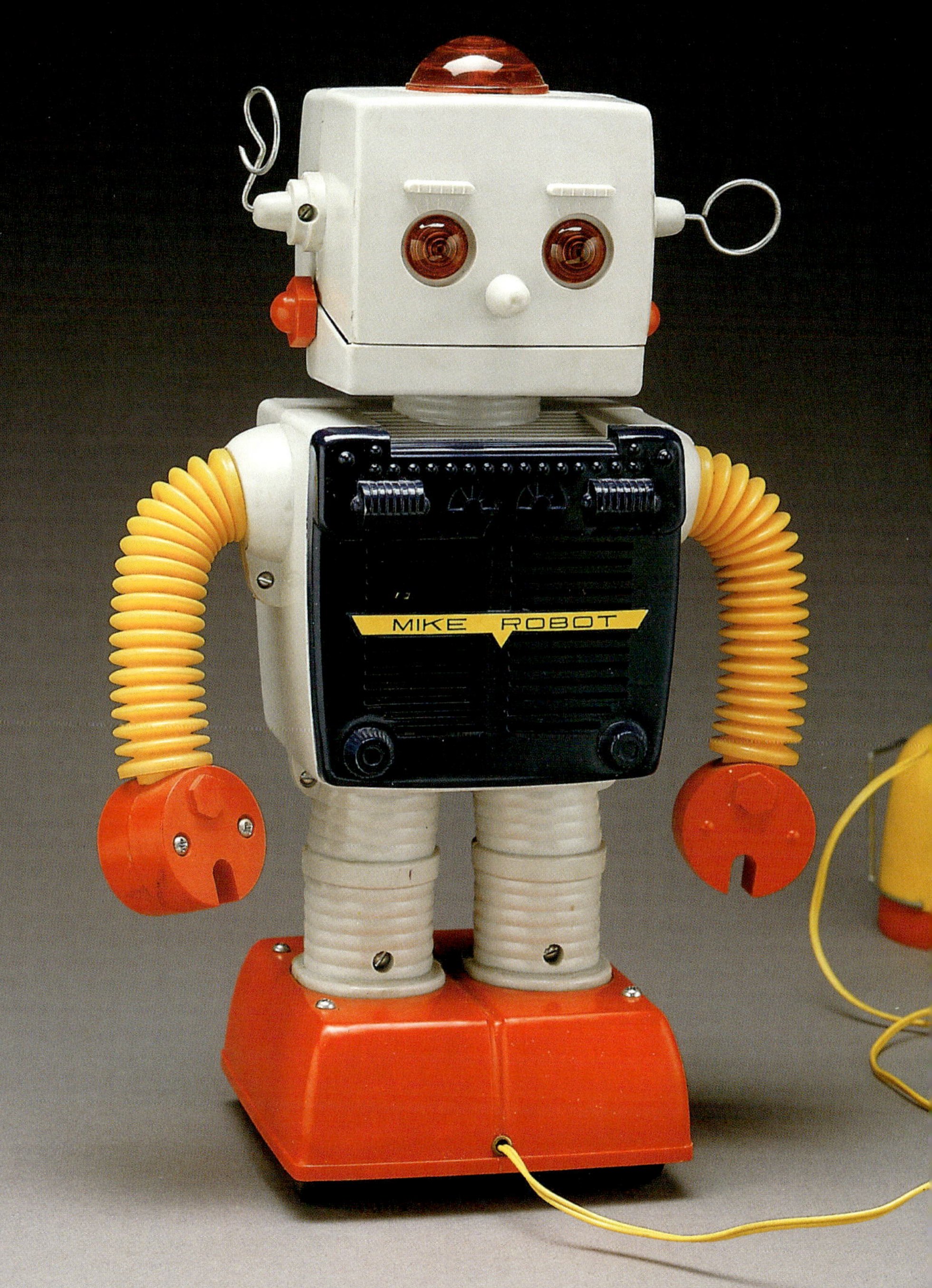
MIKE ROBOT

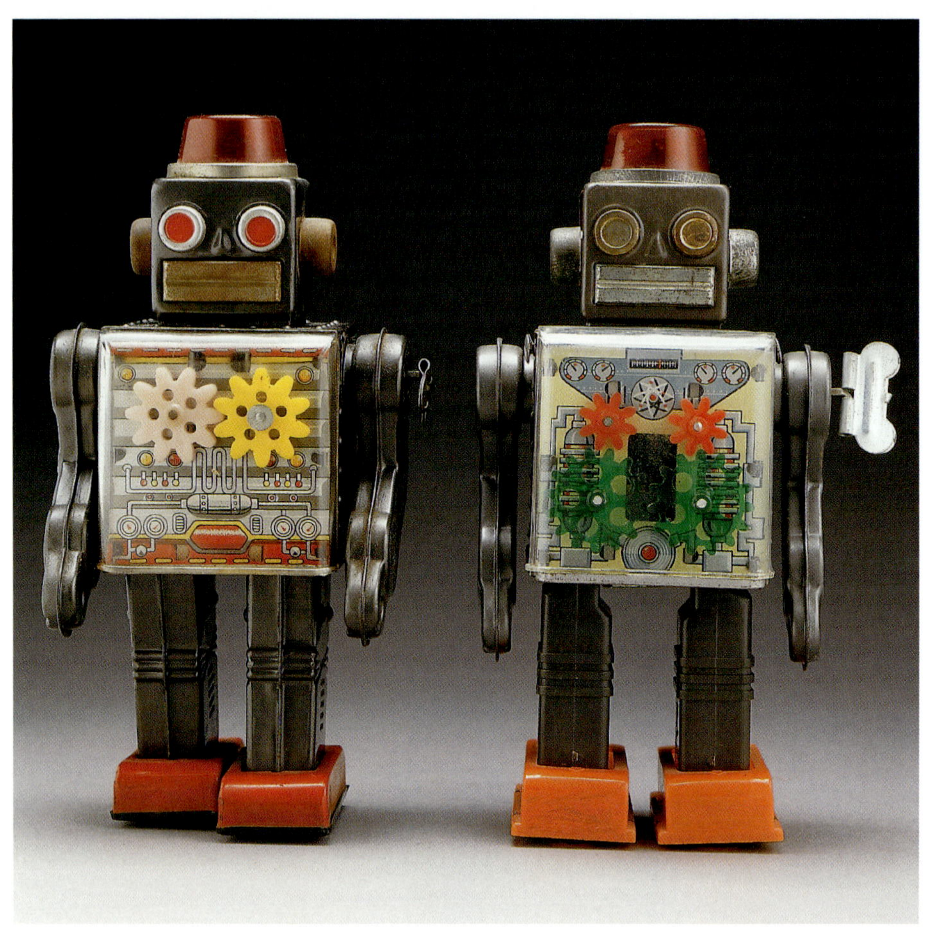

1960s, *Gear Robots*, Horikawa, 65 x 115 x 210 mm

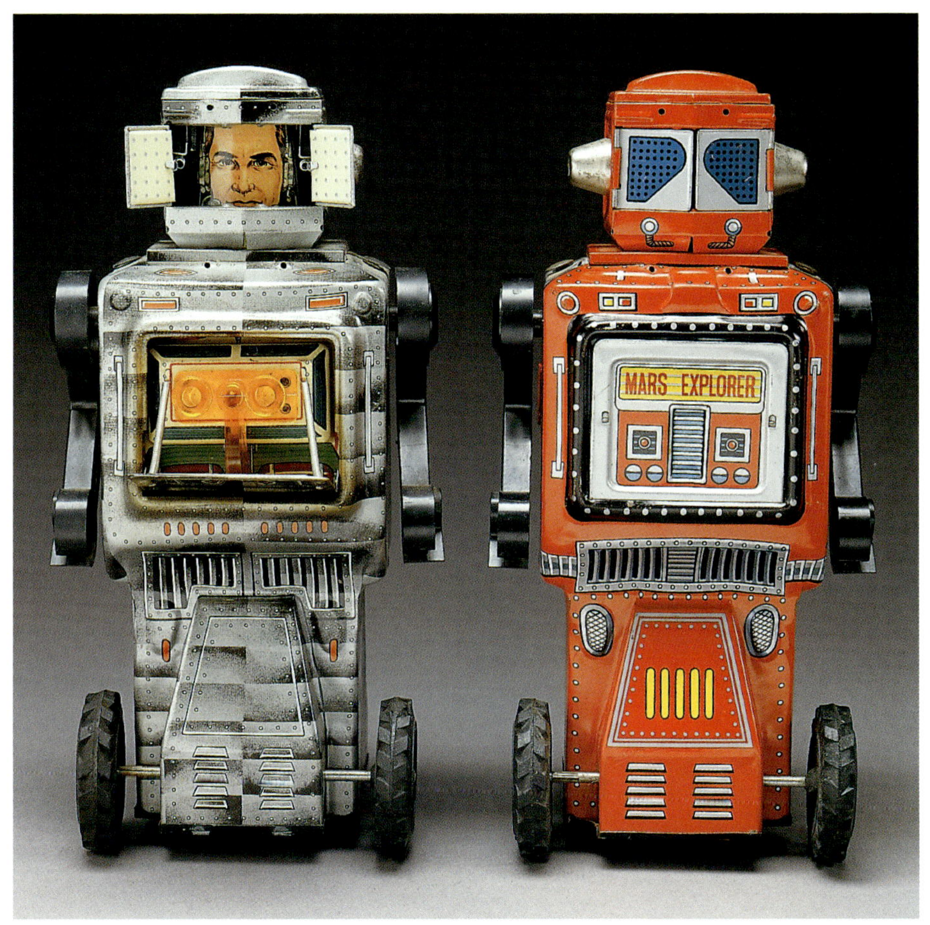

1960s, *Mars Explorers*, Horikawa, 105 x 120 x 240 mm

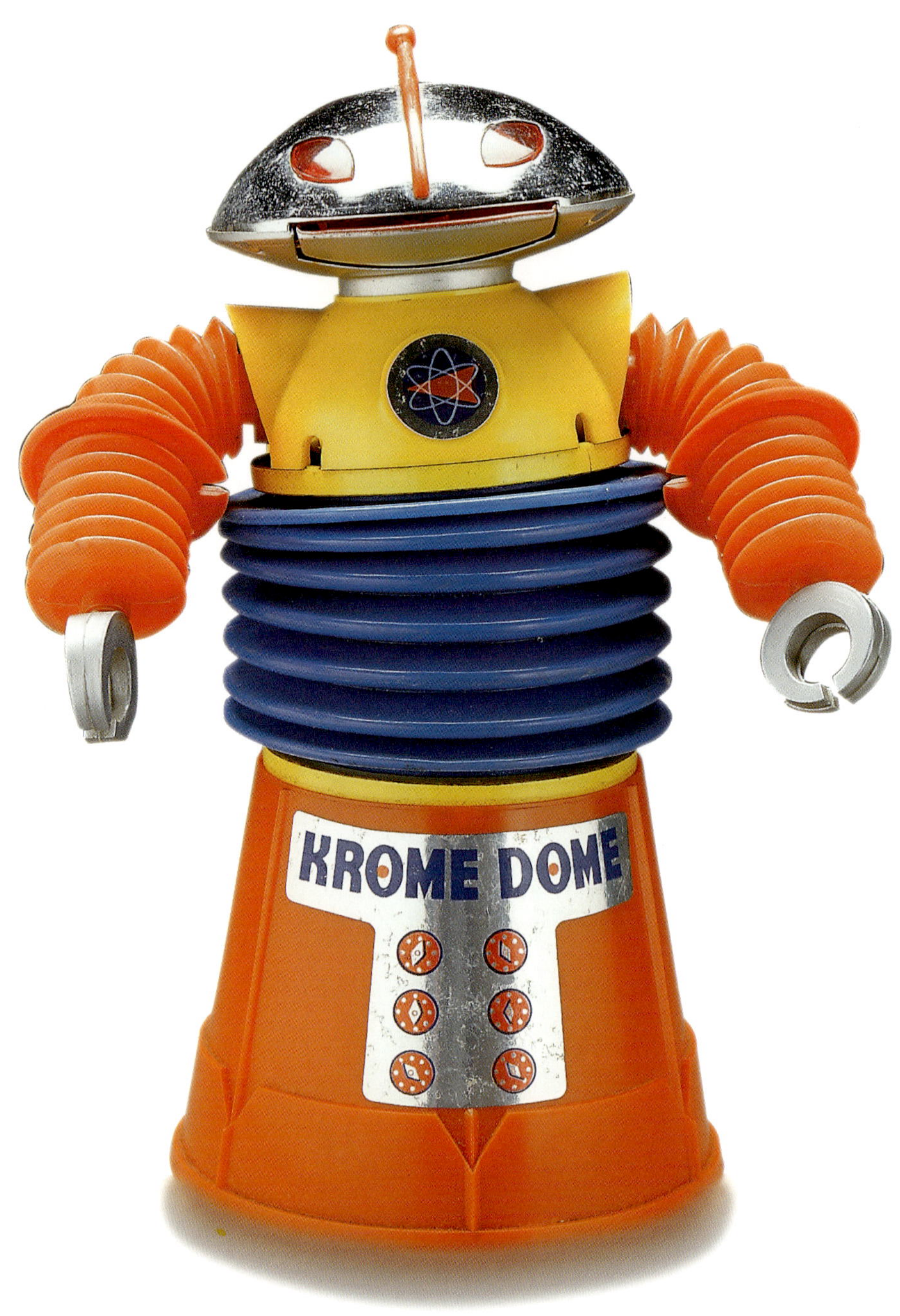

1960s, *Krome Dome*, Yonezawa, 130 x 130 x 260 mm

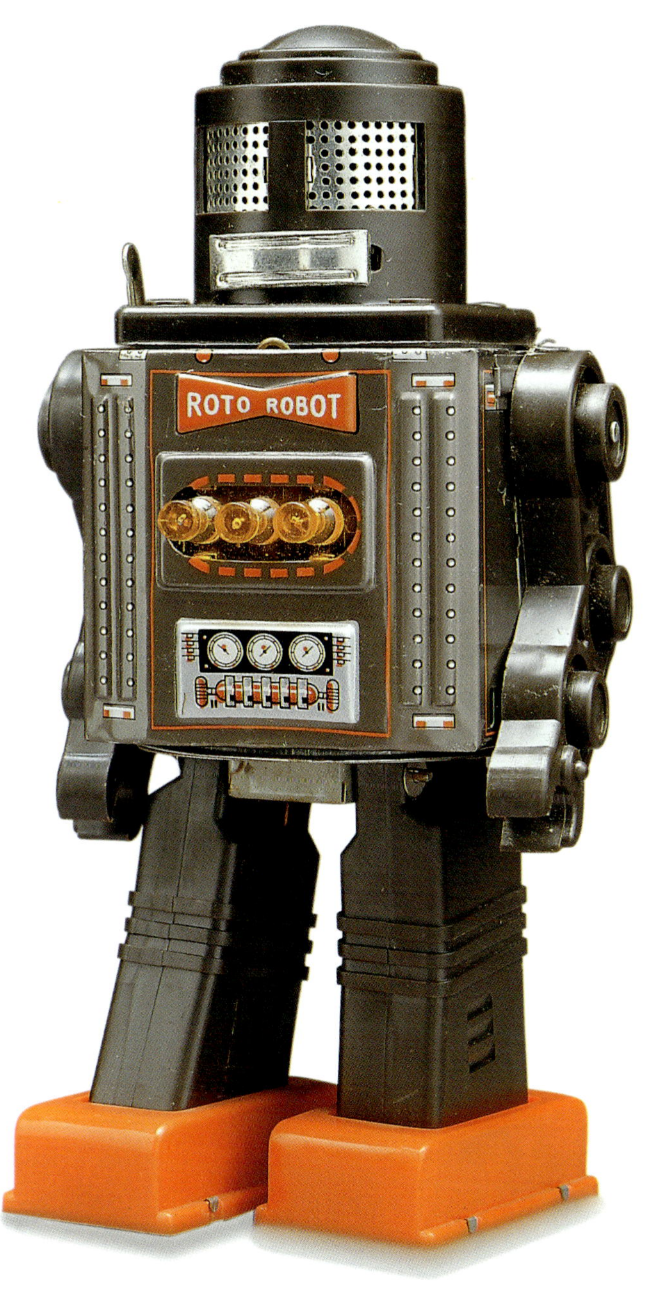

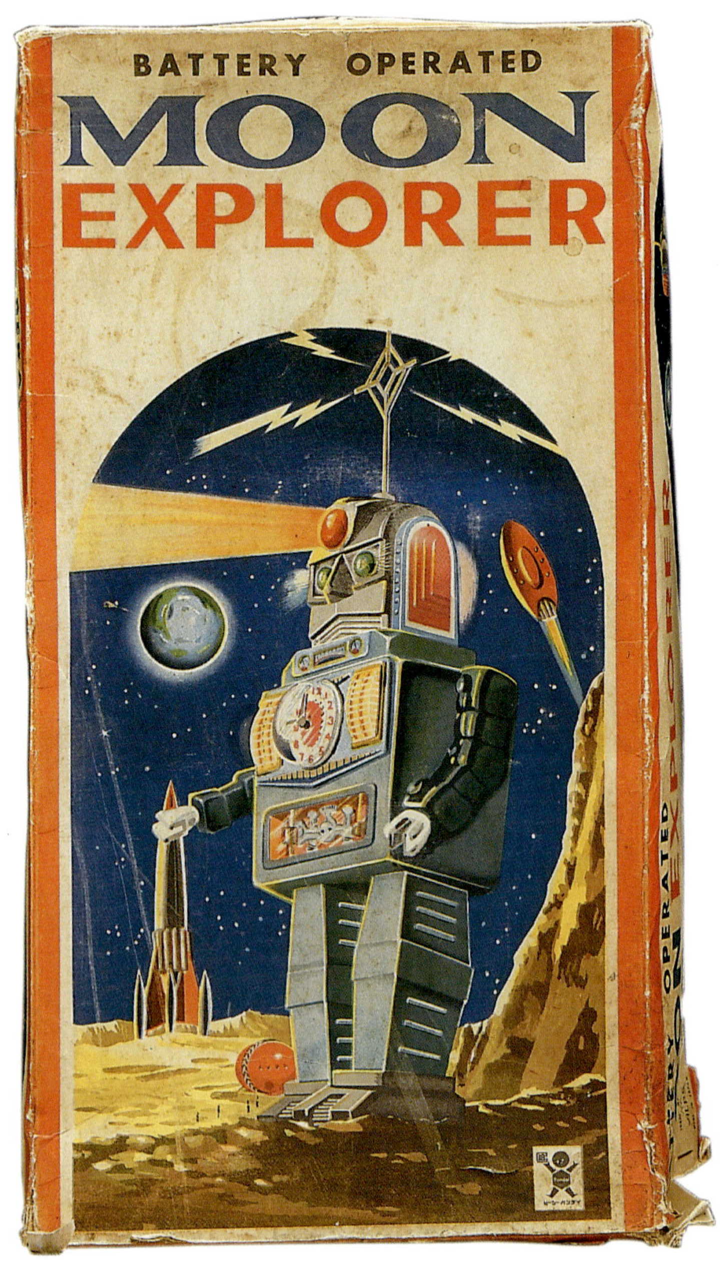

1960s, *Moon Explorer*, Bandai, 107 x 165 x 452 mm

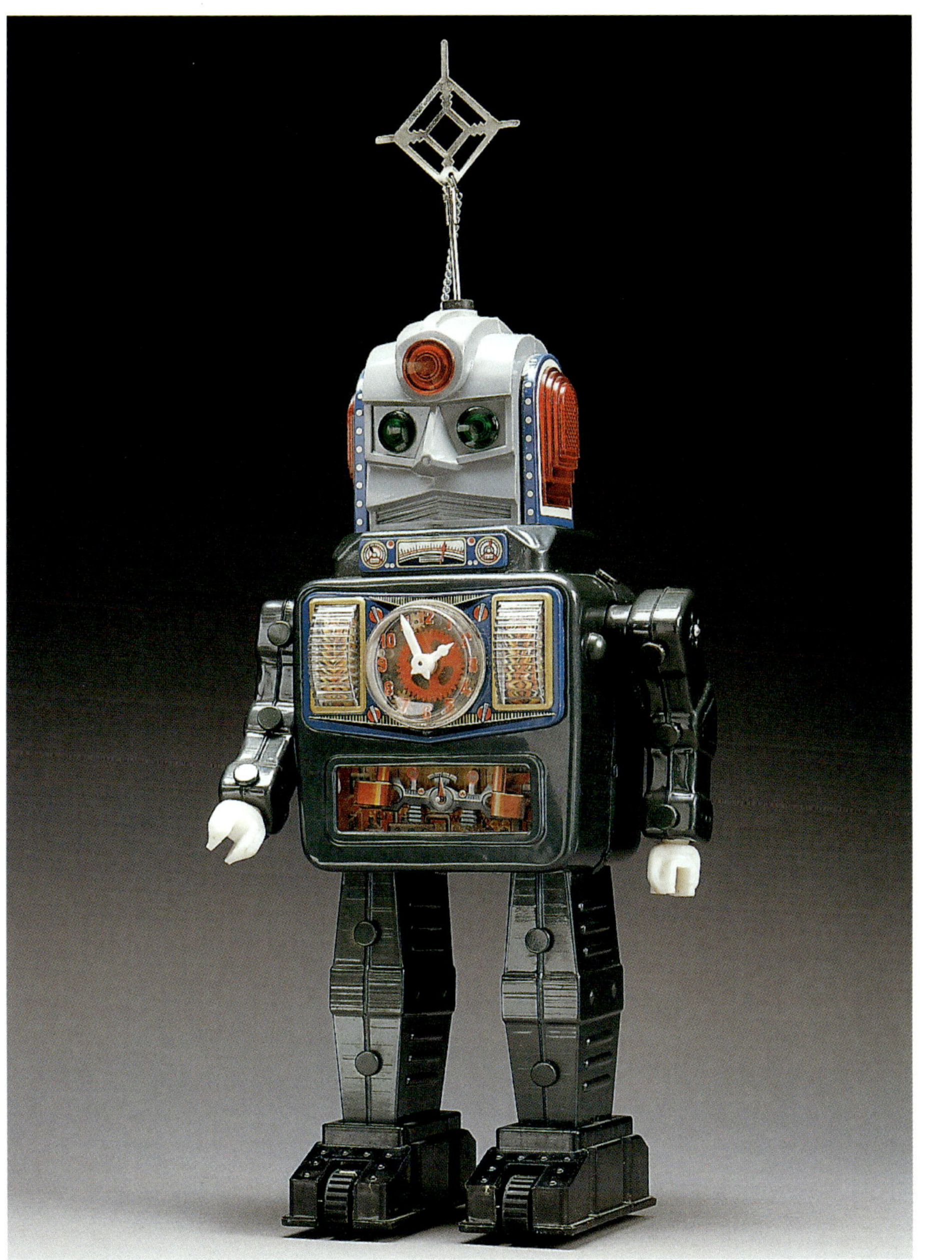

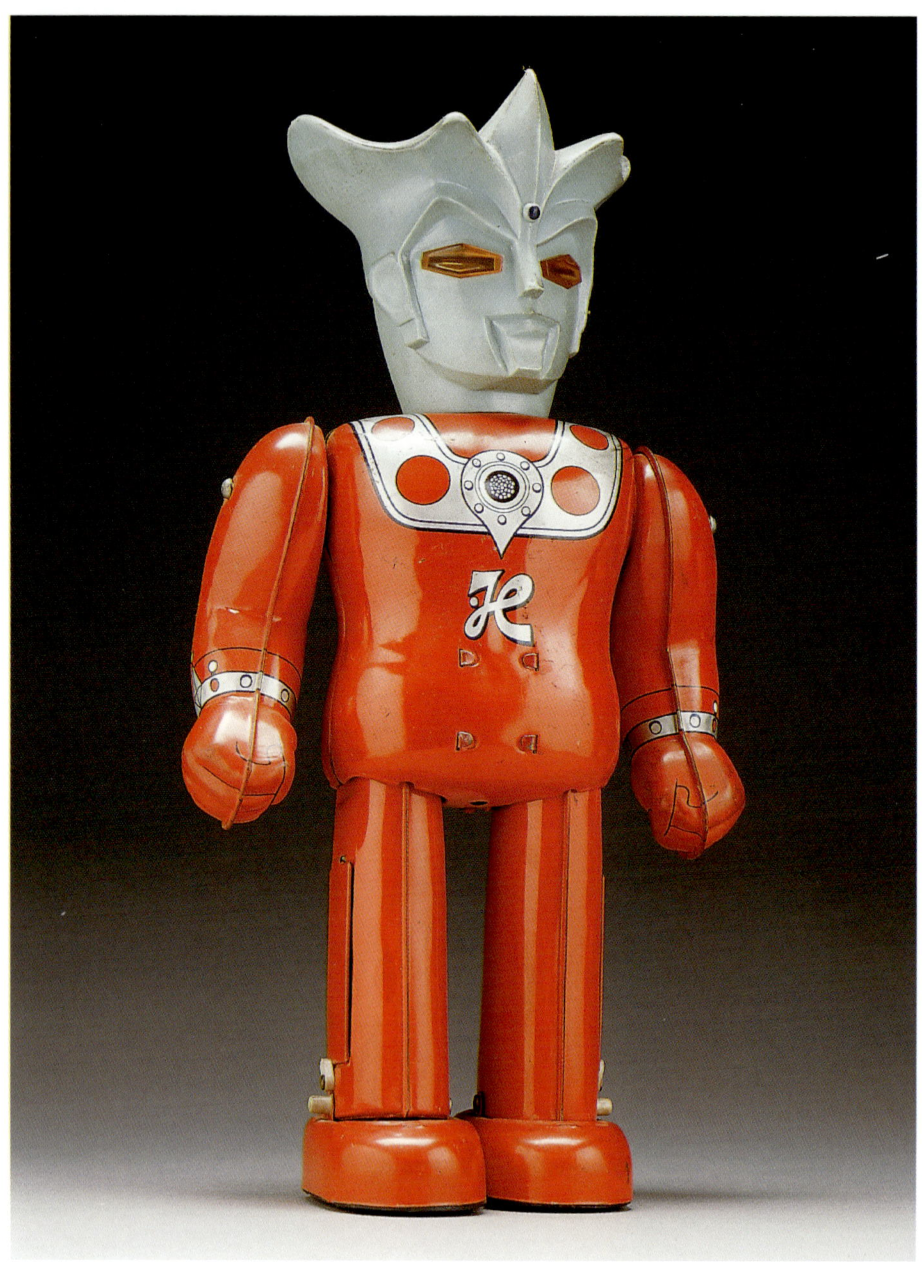

1970s, *Ultra Man Leo*, Bull Mark, 85 x 160 x 320 mm

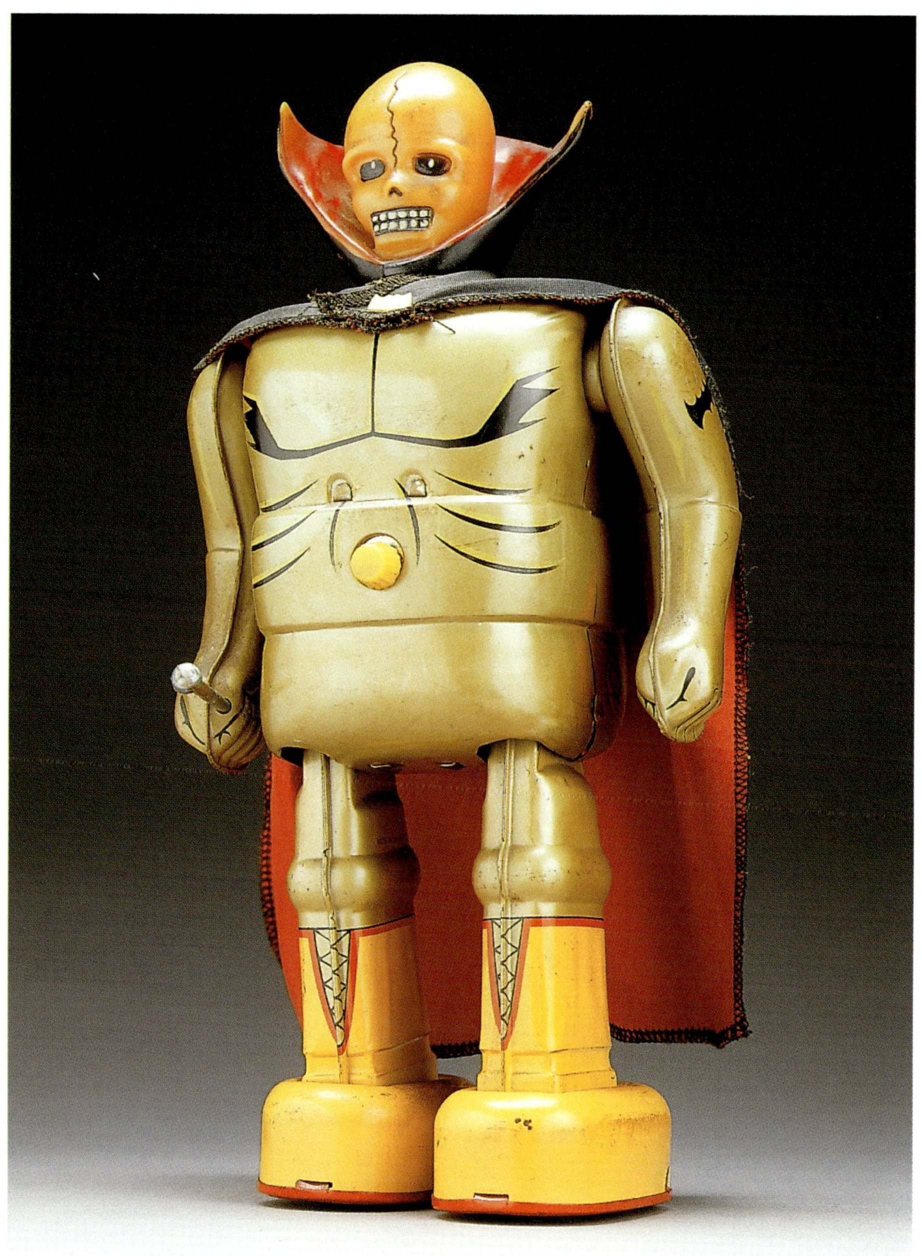

1960s, *Ohgon Bat*, Nomura, 90 x 135 x 285 mm

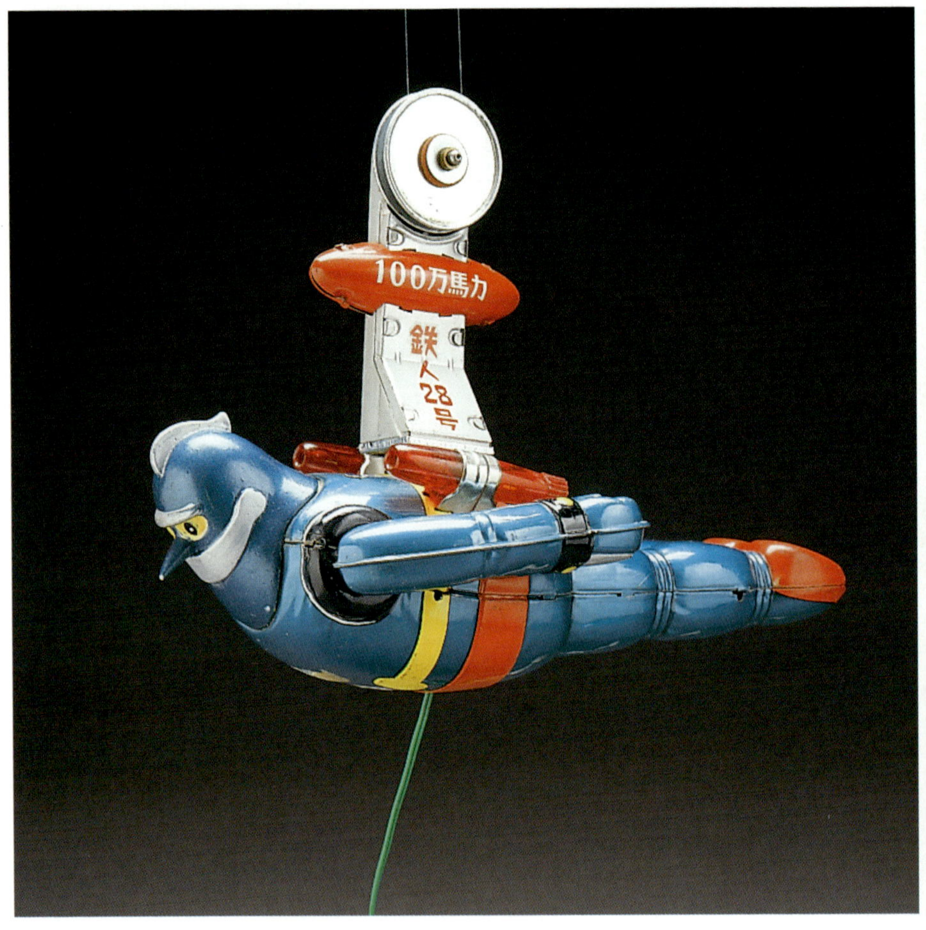

▲ 1960s, *Tetsujin 28-GO*, Nomura, 230 x 240 x 90 mm

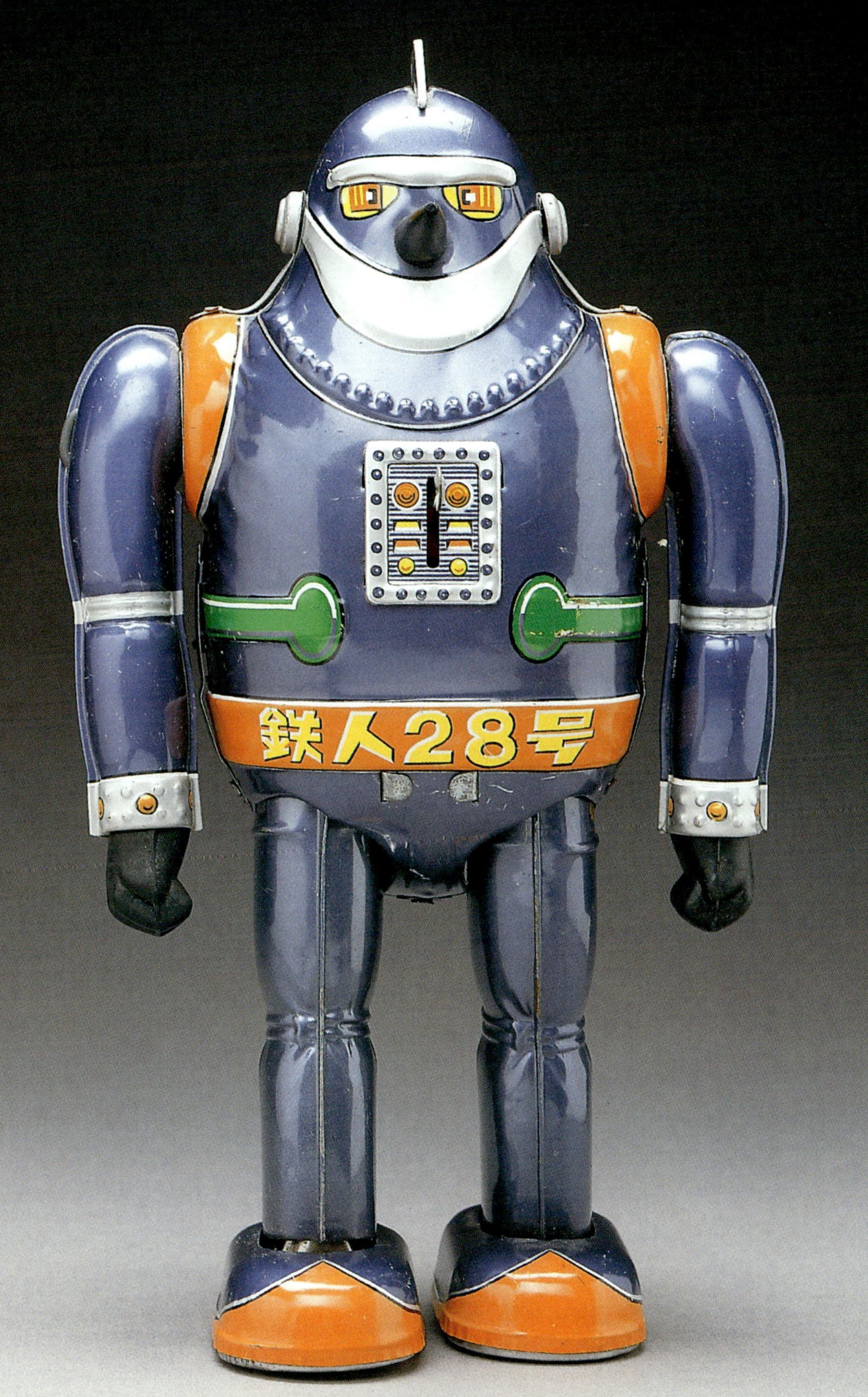
鉄人28号

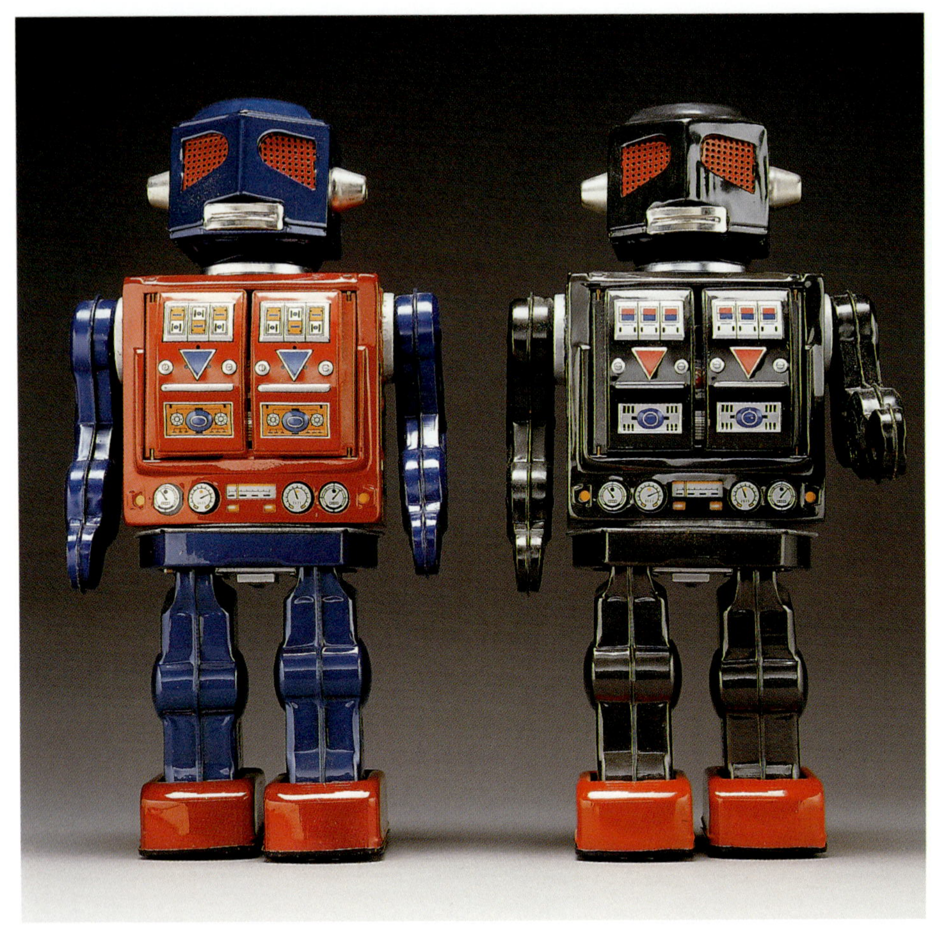

1960s, *Robots*, Horikawa, 100 x 140 x 290 mm

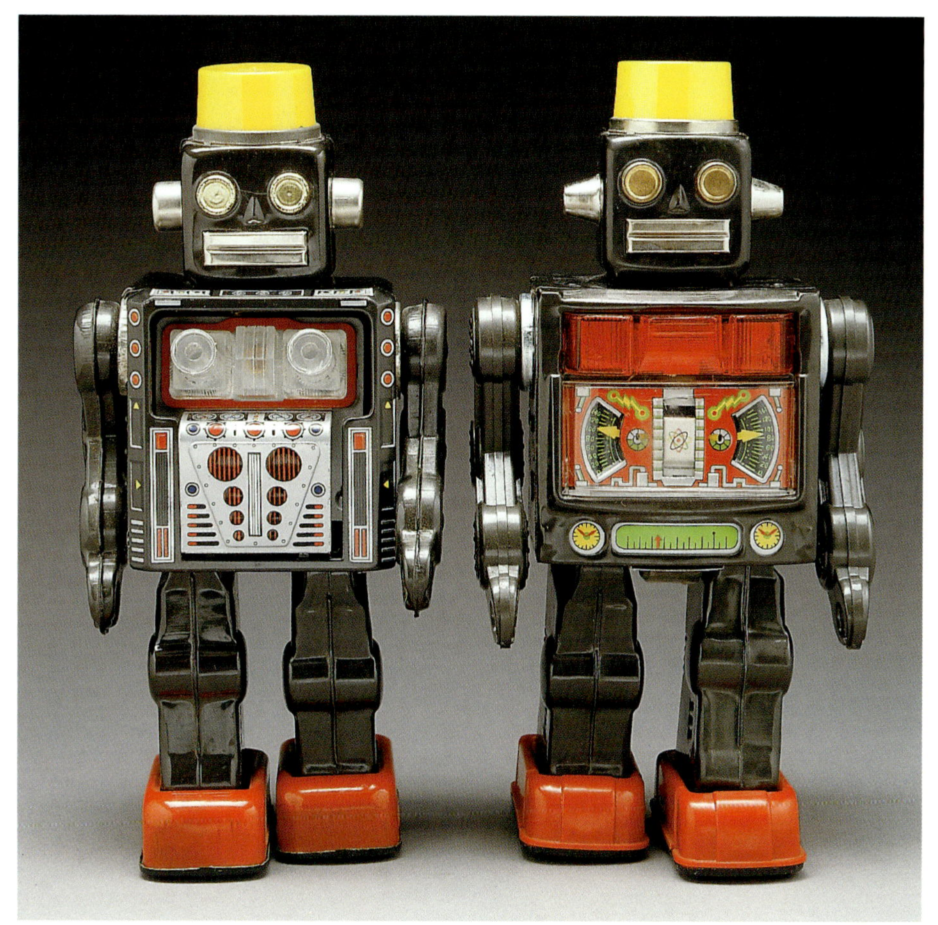

1960s, *Thunder Robots*, Horikawa, 90 x 135 x 285 mm

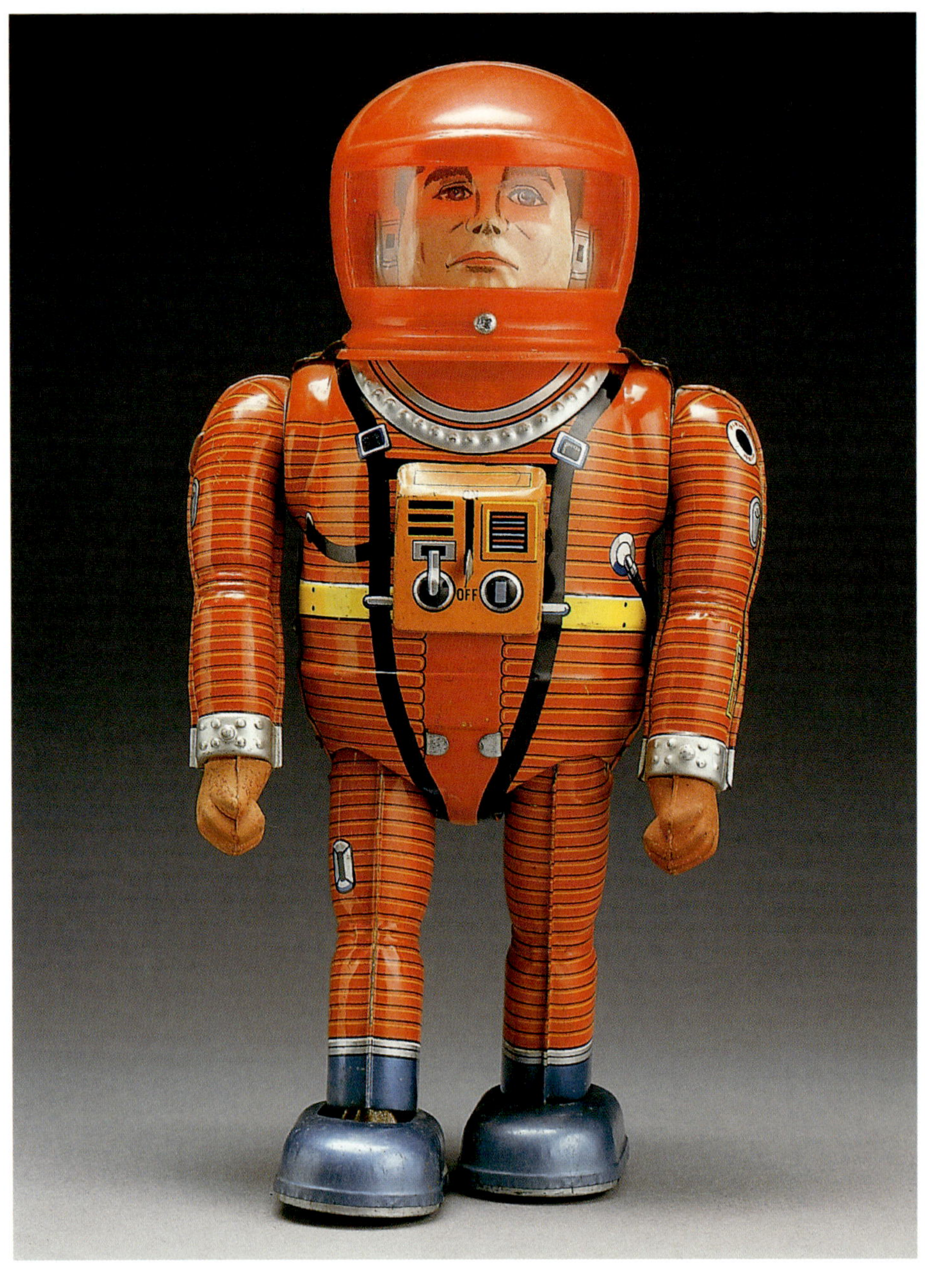

1960s, *Astro Man*, Nomura, 90 x 135 x 260 mm

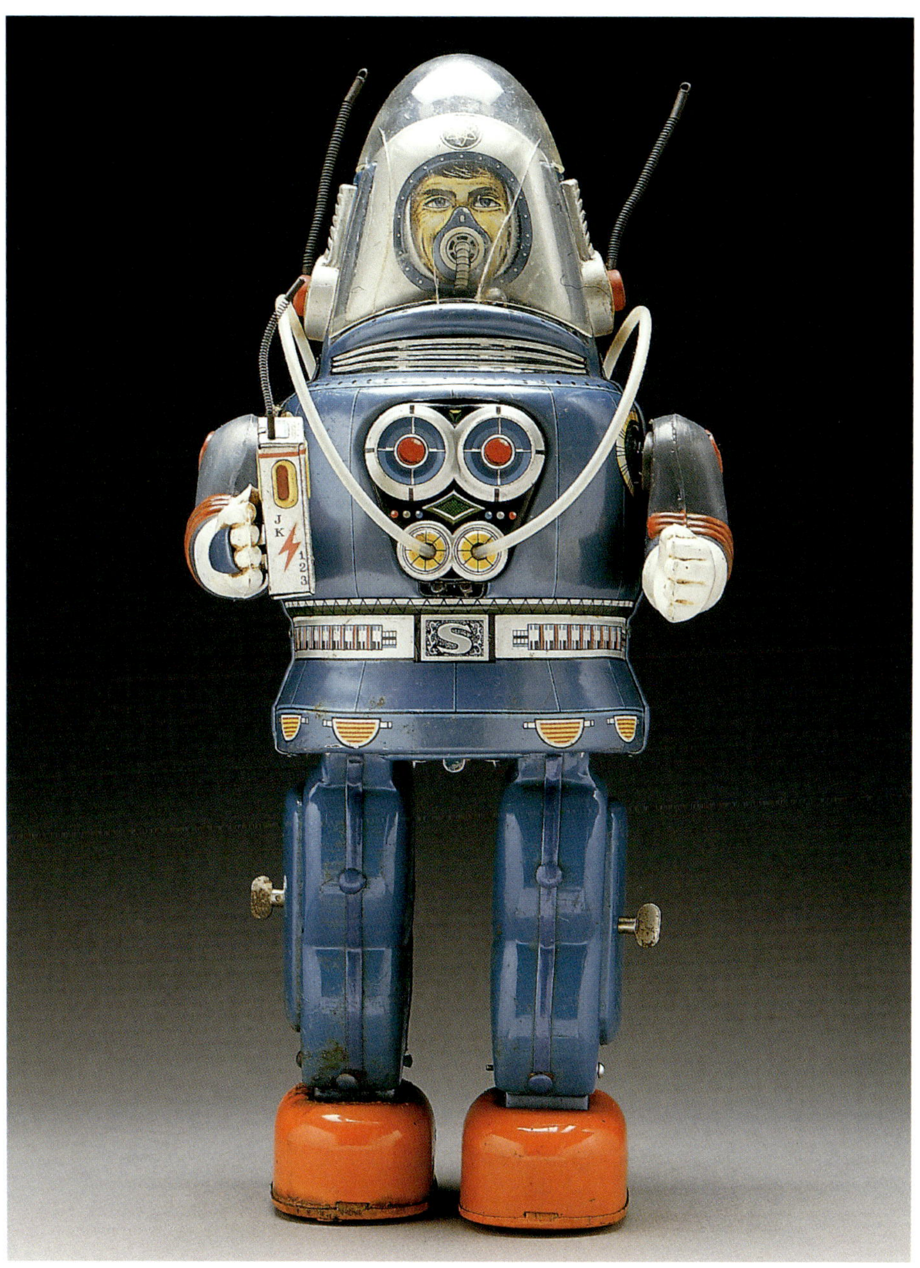

1960s, *Astronaut*, Rosko Toy, 110 x 135 x 330 mm

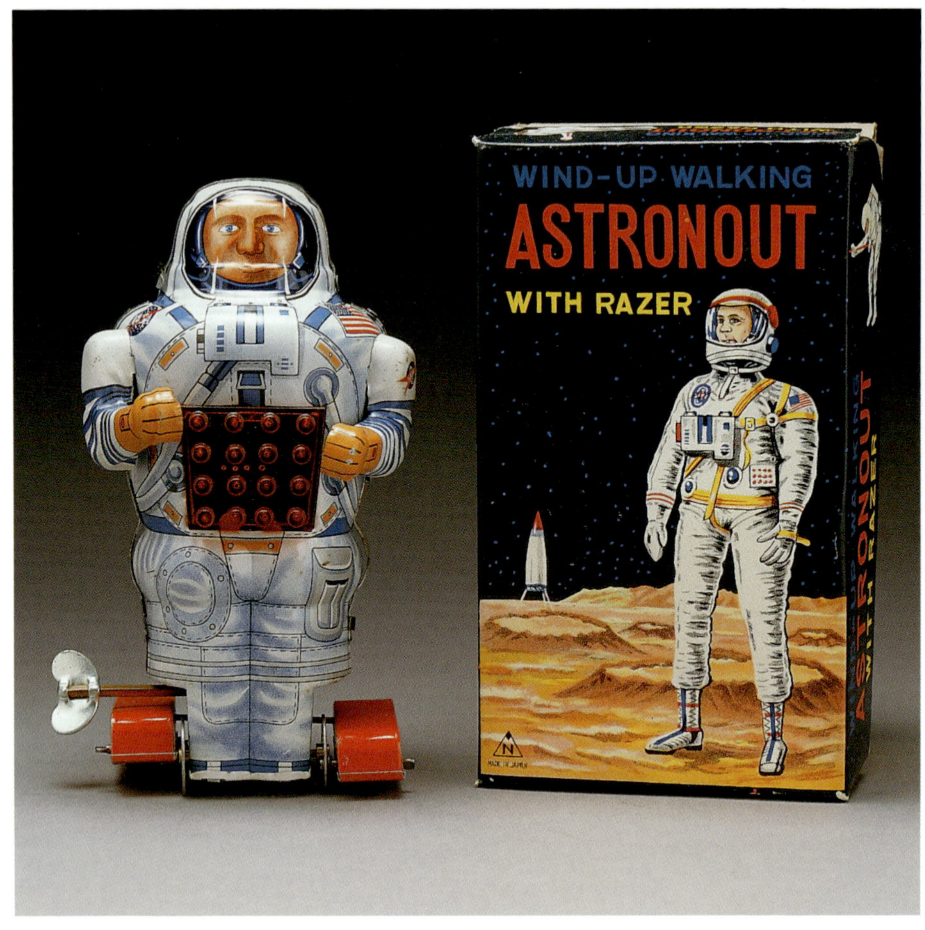

1970s, *Astronaut and Case*, unknown, 85 x 95 x 155 mm

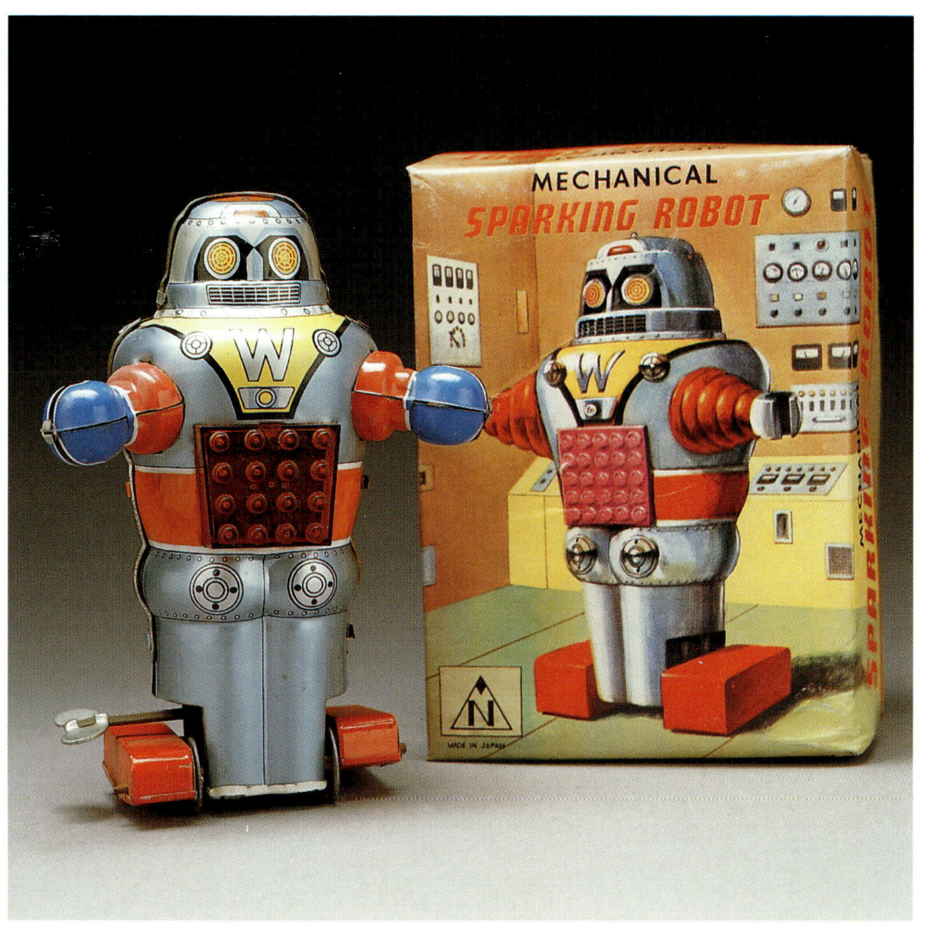

1950s, *Sparking Robot and Case*, unknown, 75 x 115 x 160 mm

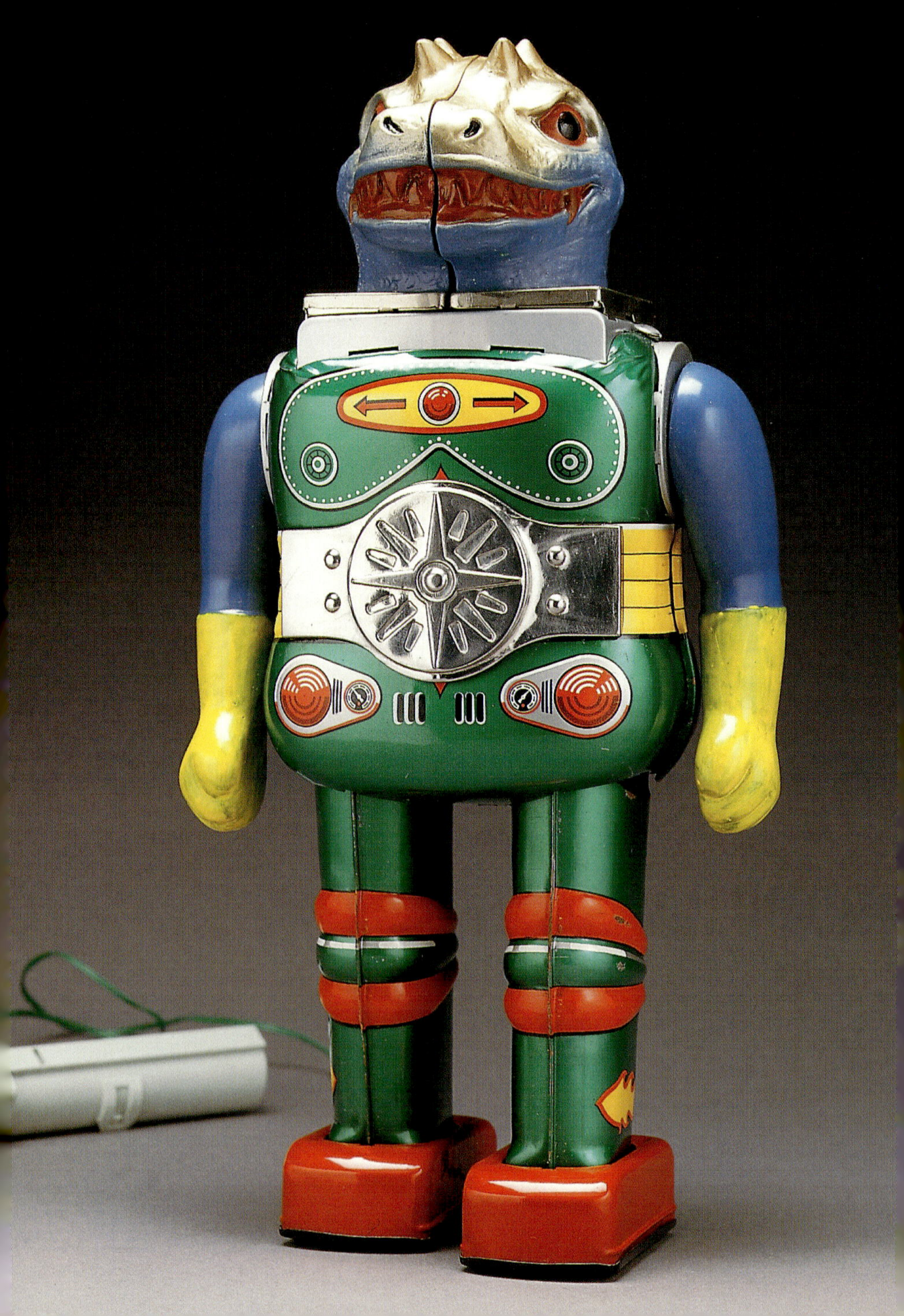

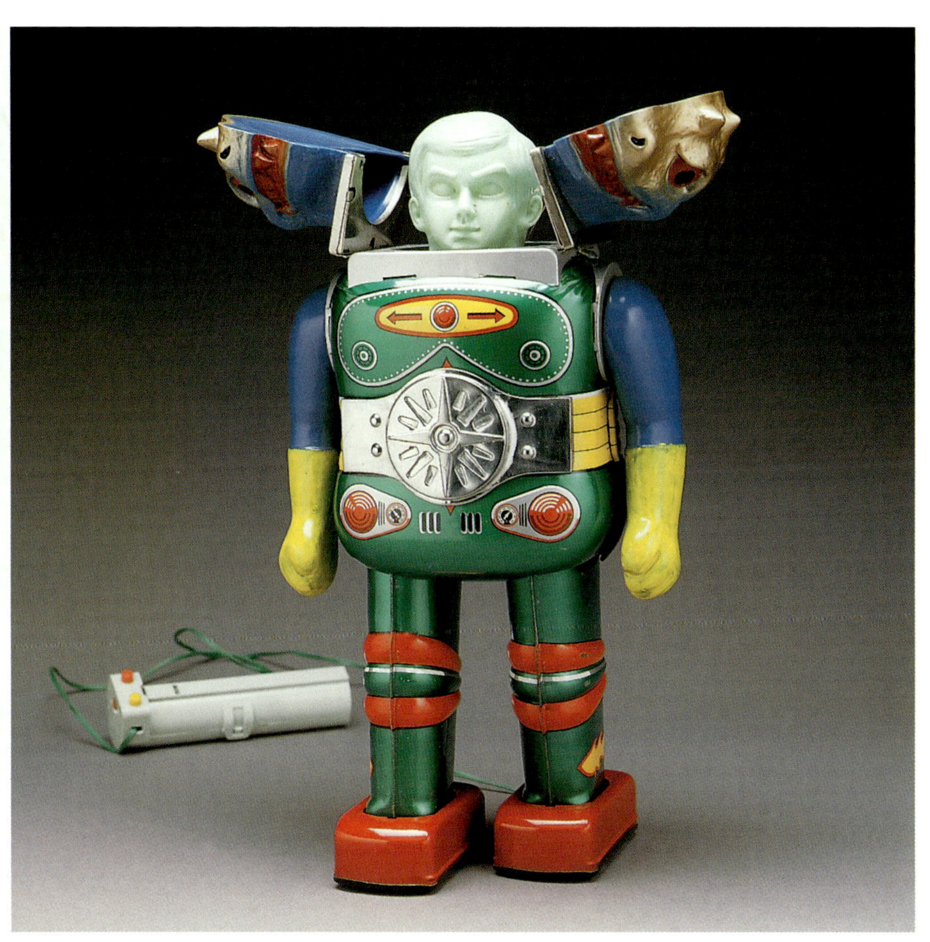

1960s, *Metamorph*, Marumiya, 100 x 160 x 335 mm

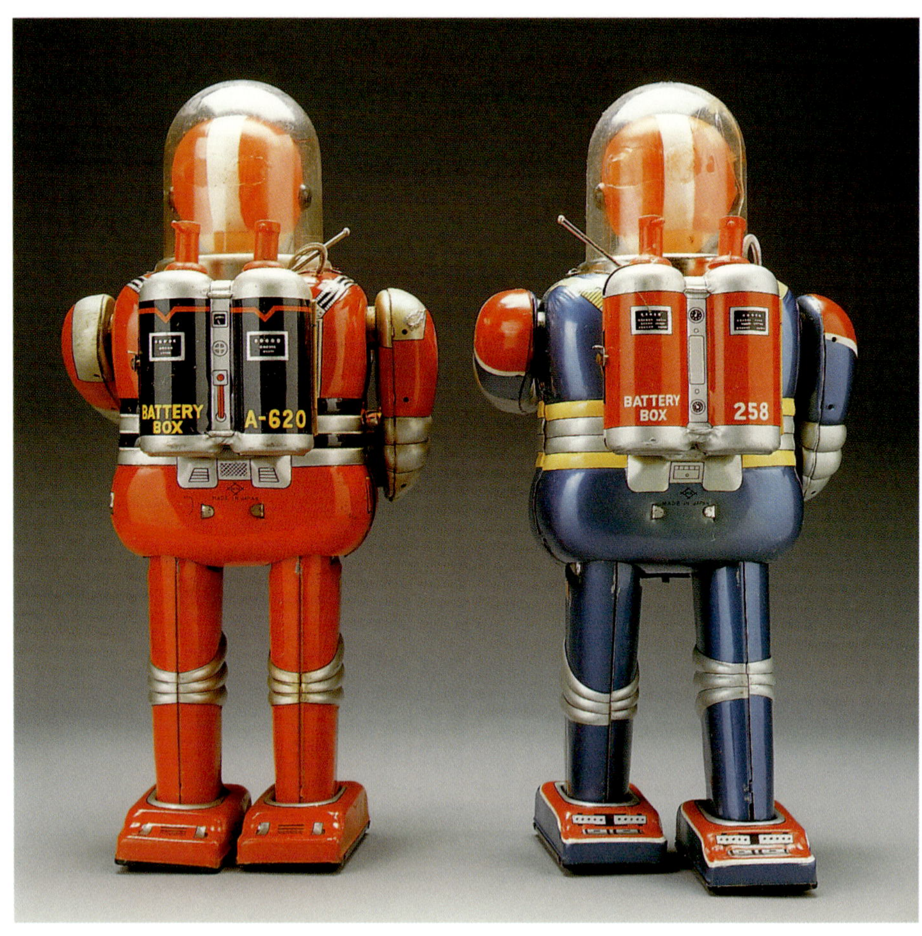

1960s, *Astronauts*, Daiya, 125 x 170 x 340 mm

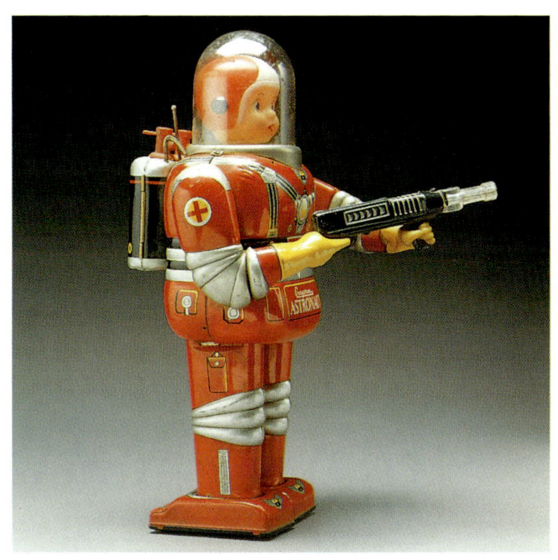

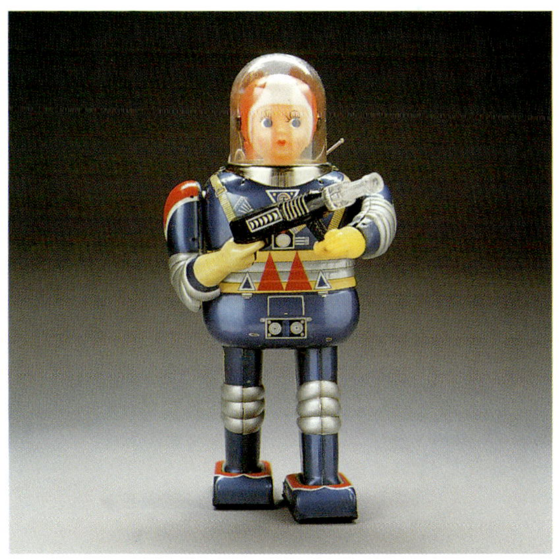

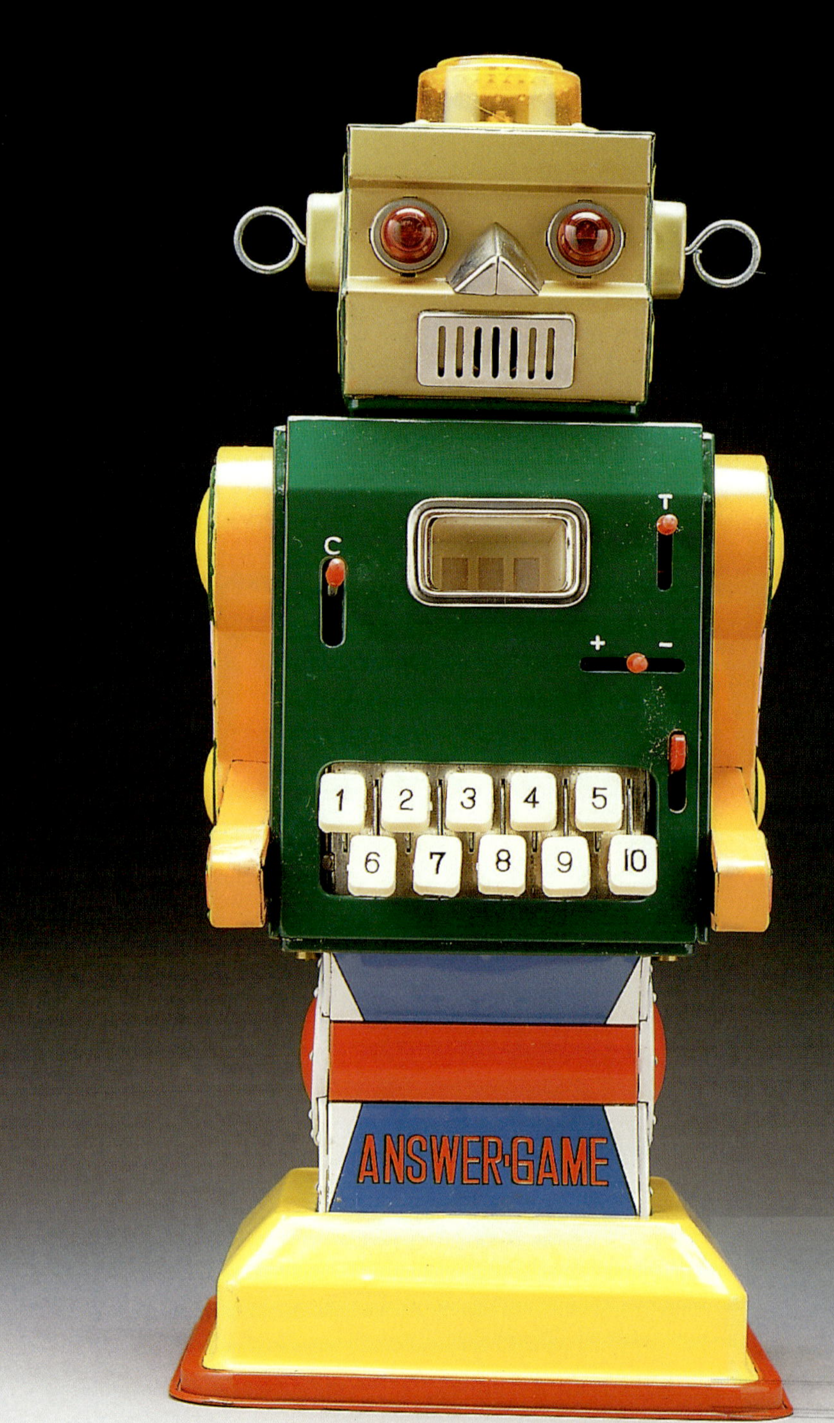

1 2 3 4 5
6 7 8 9 10
C
T
+ −
ANSWER·GAME

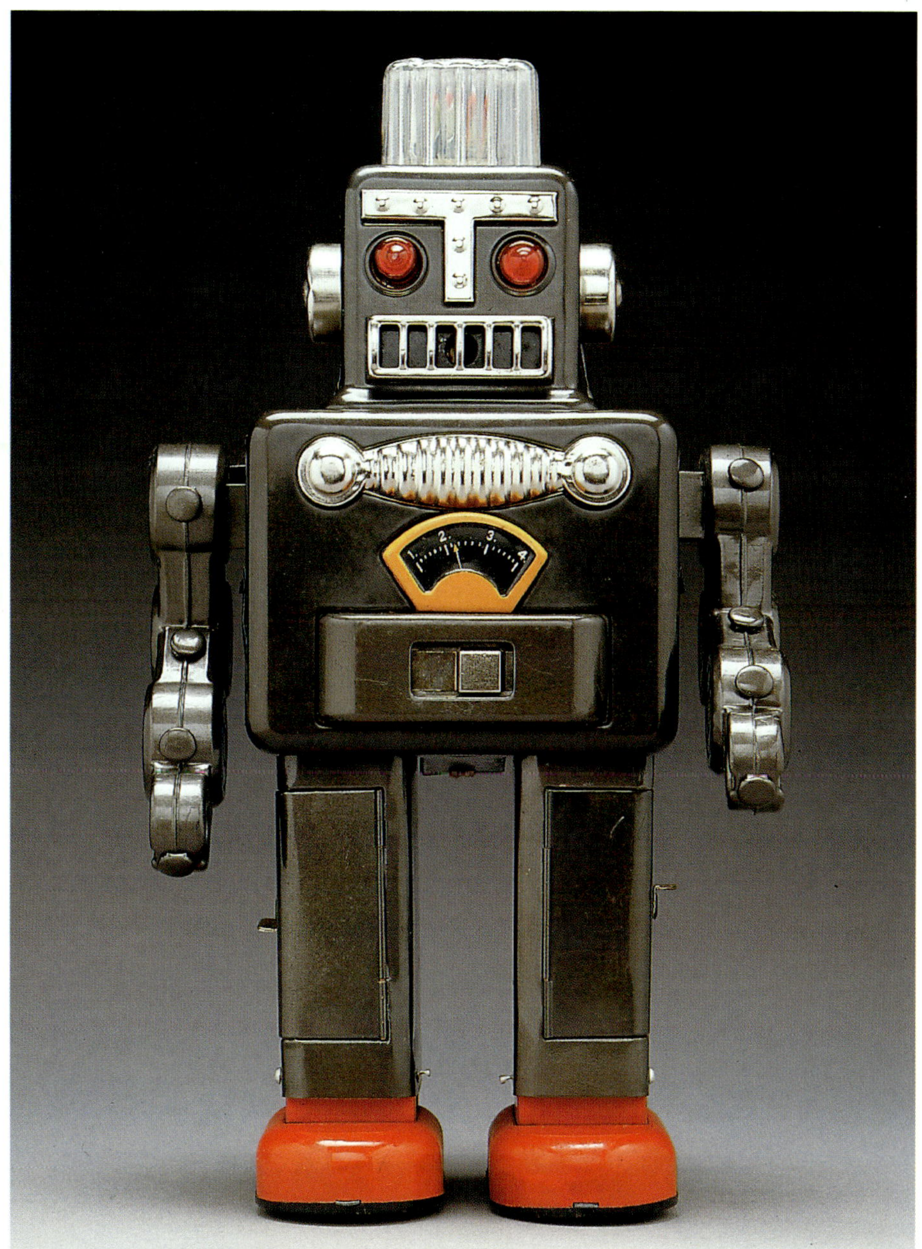

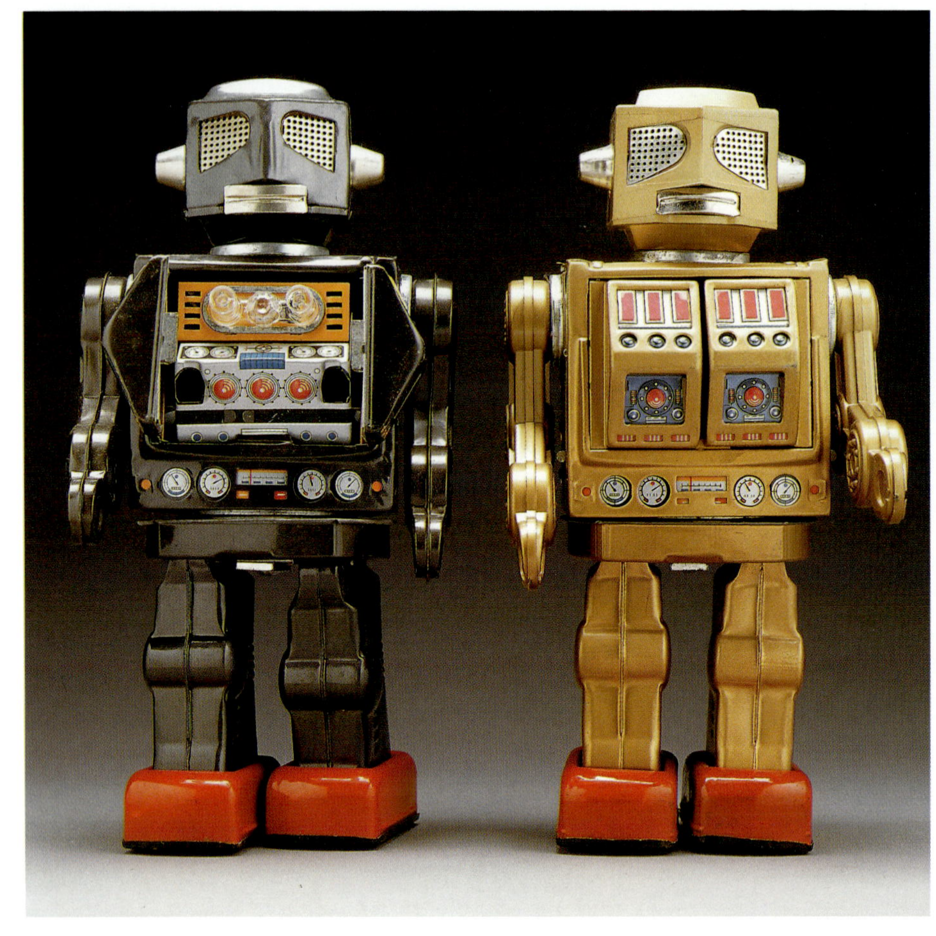

1960s, *Robots*, Horikawa, 105 x 140 x 290 mm

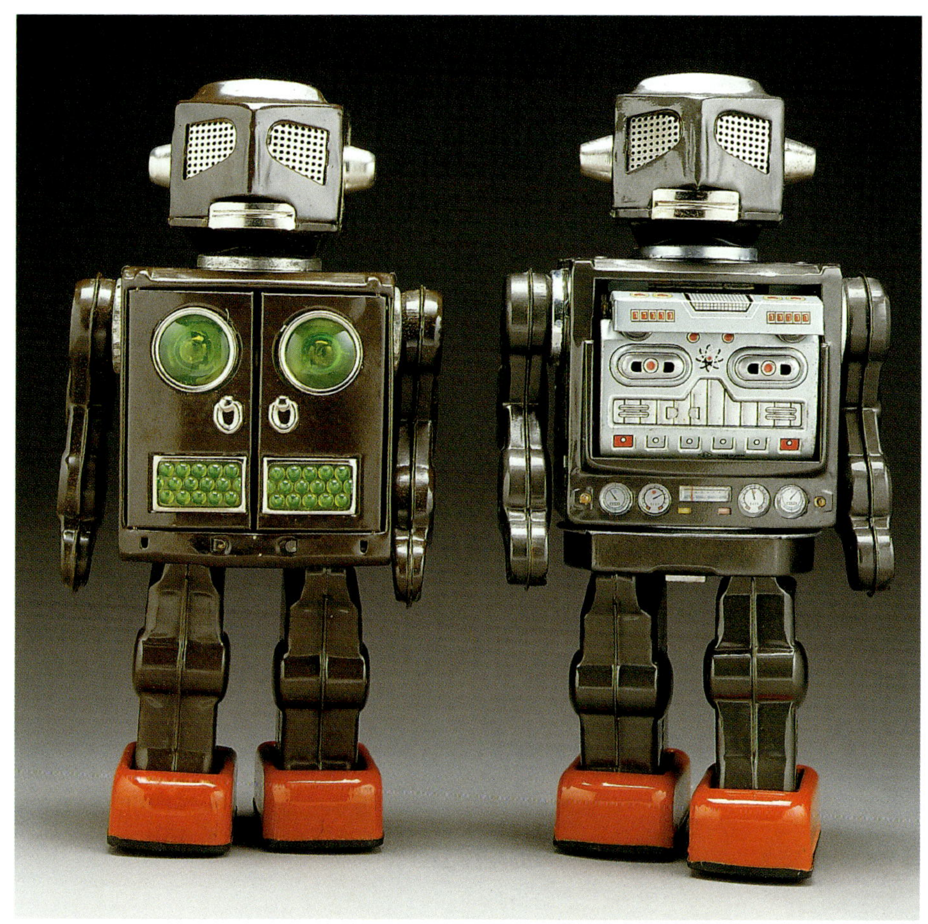

1960s, *Robots*, Horikawa, 87 x 135 x 280, 87 x 135 x 290 mm

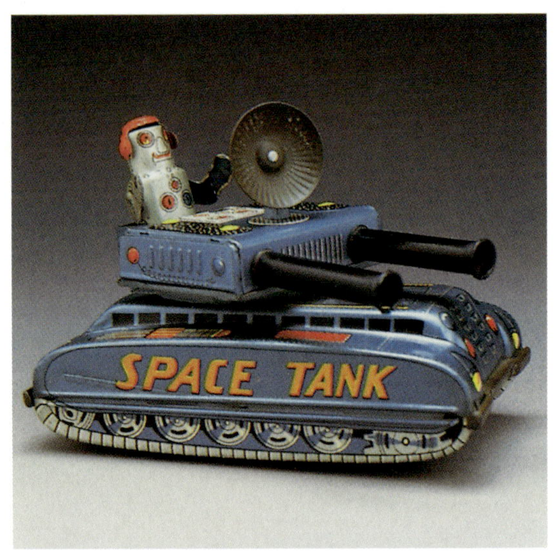

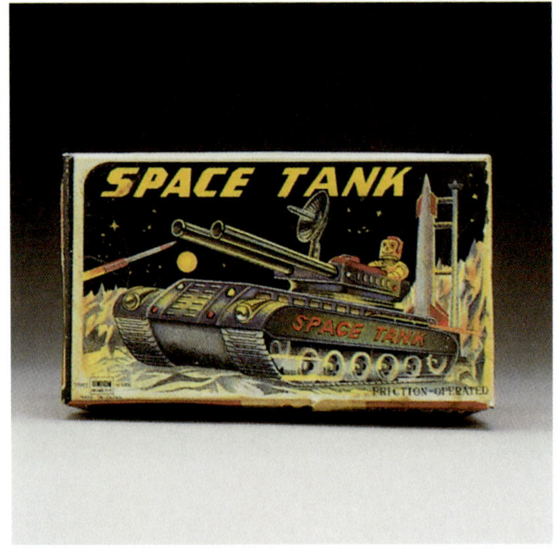

1950s, *Space Tank*, Union, 157 x 90 x 115 mm

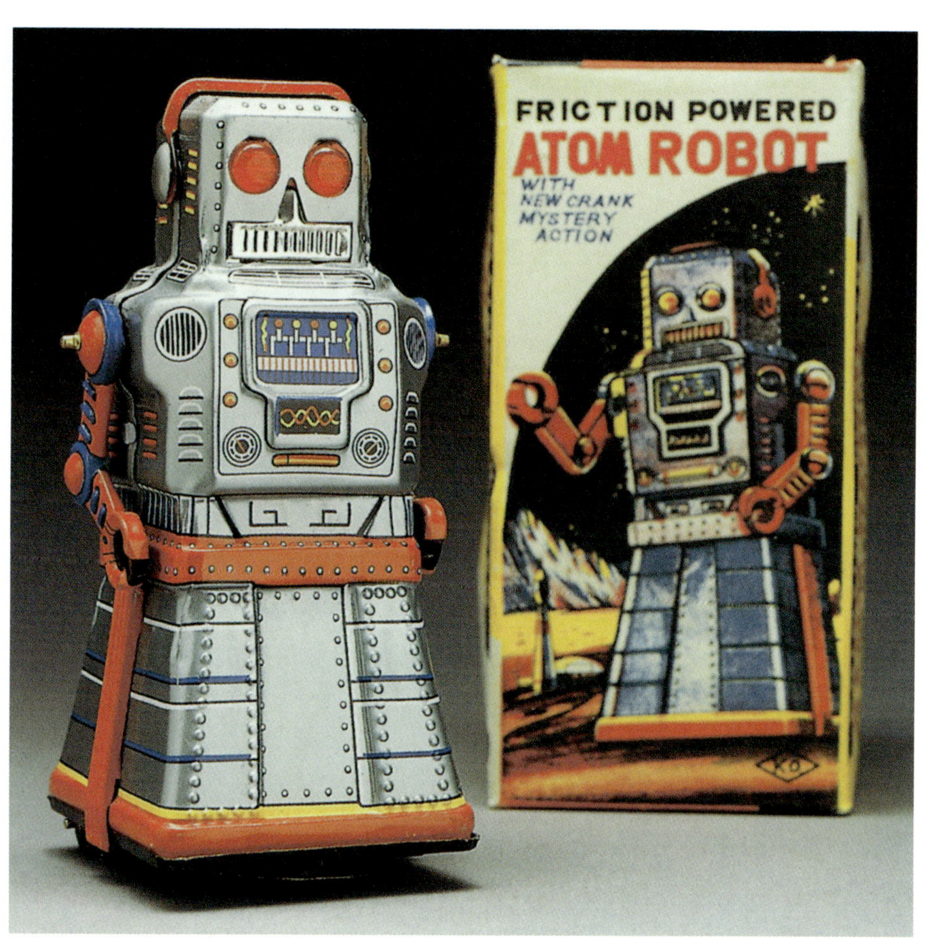

1960s, *Atom Robot and Case*, Yoshiya, 80 x 70 x 160 mm

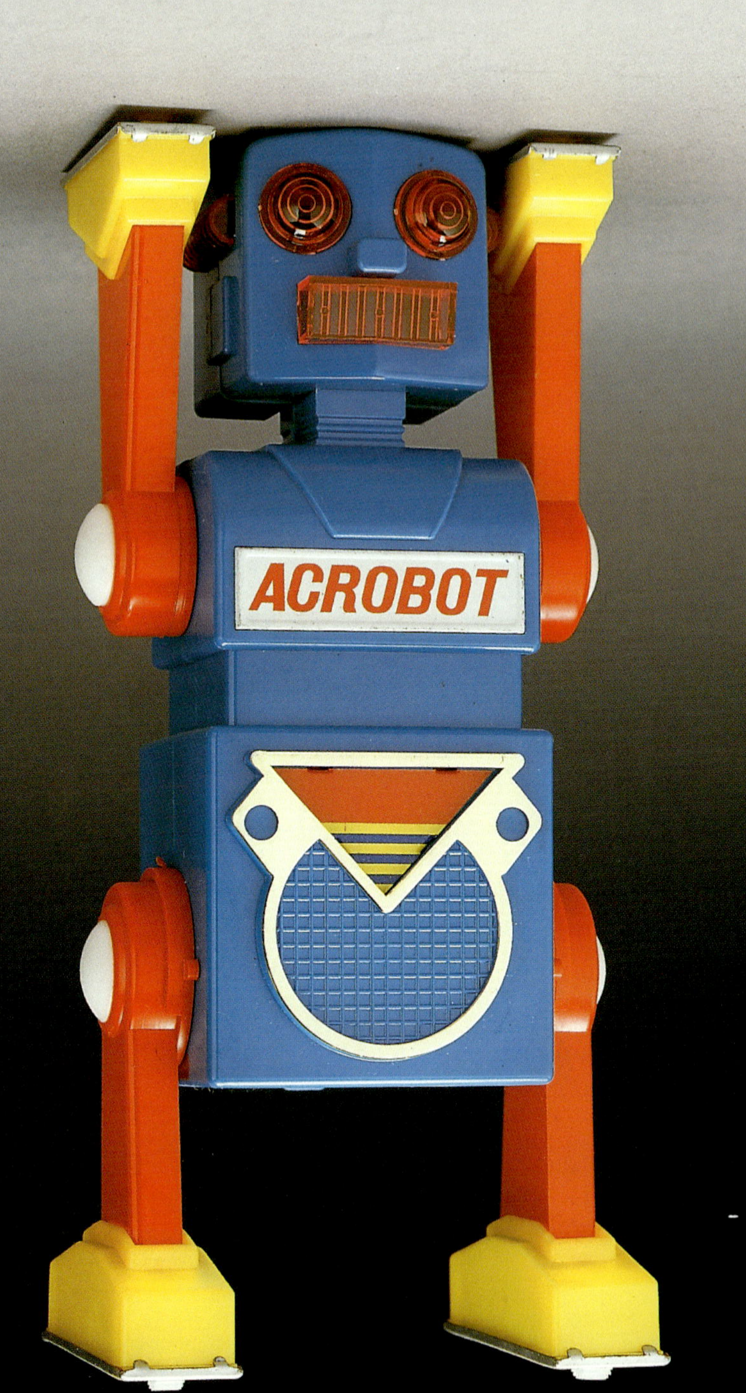
ACROBOT

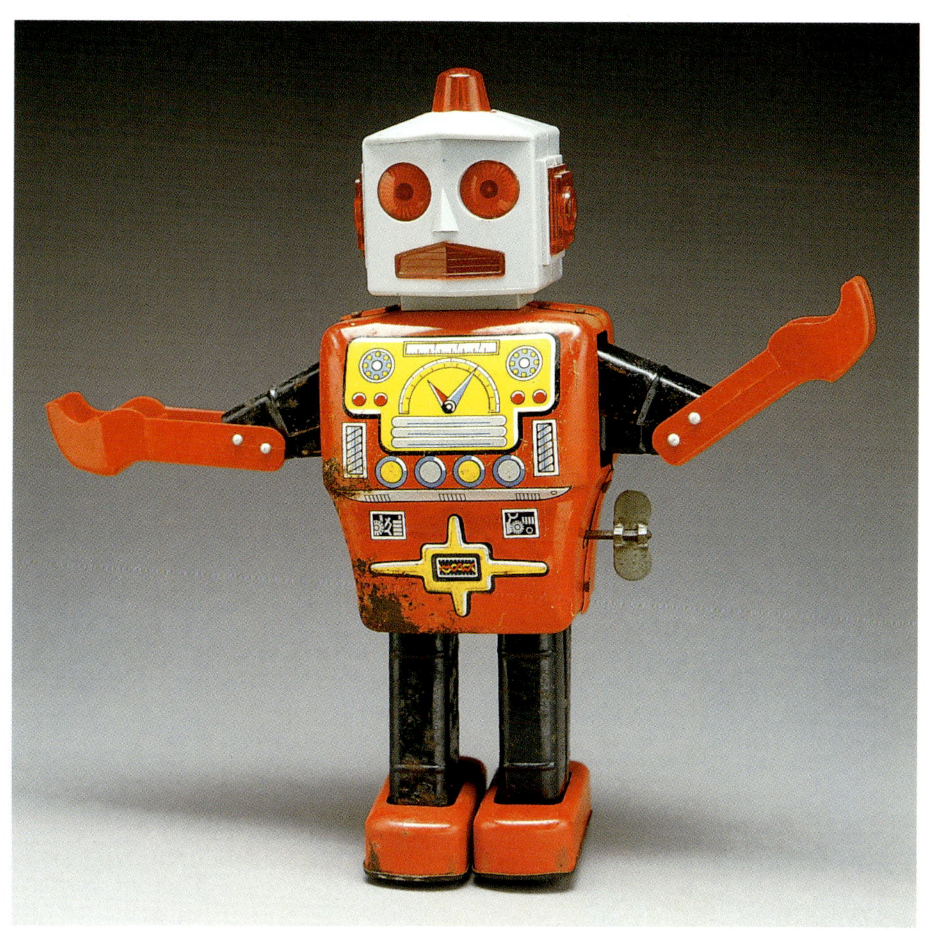

▲ 1960s, *Robot*, Yonezawa, 70 x 150 x 260 mm

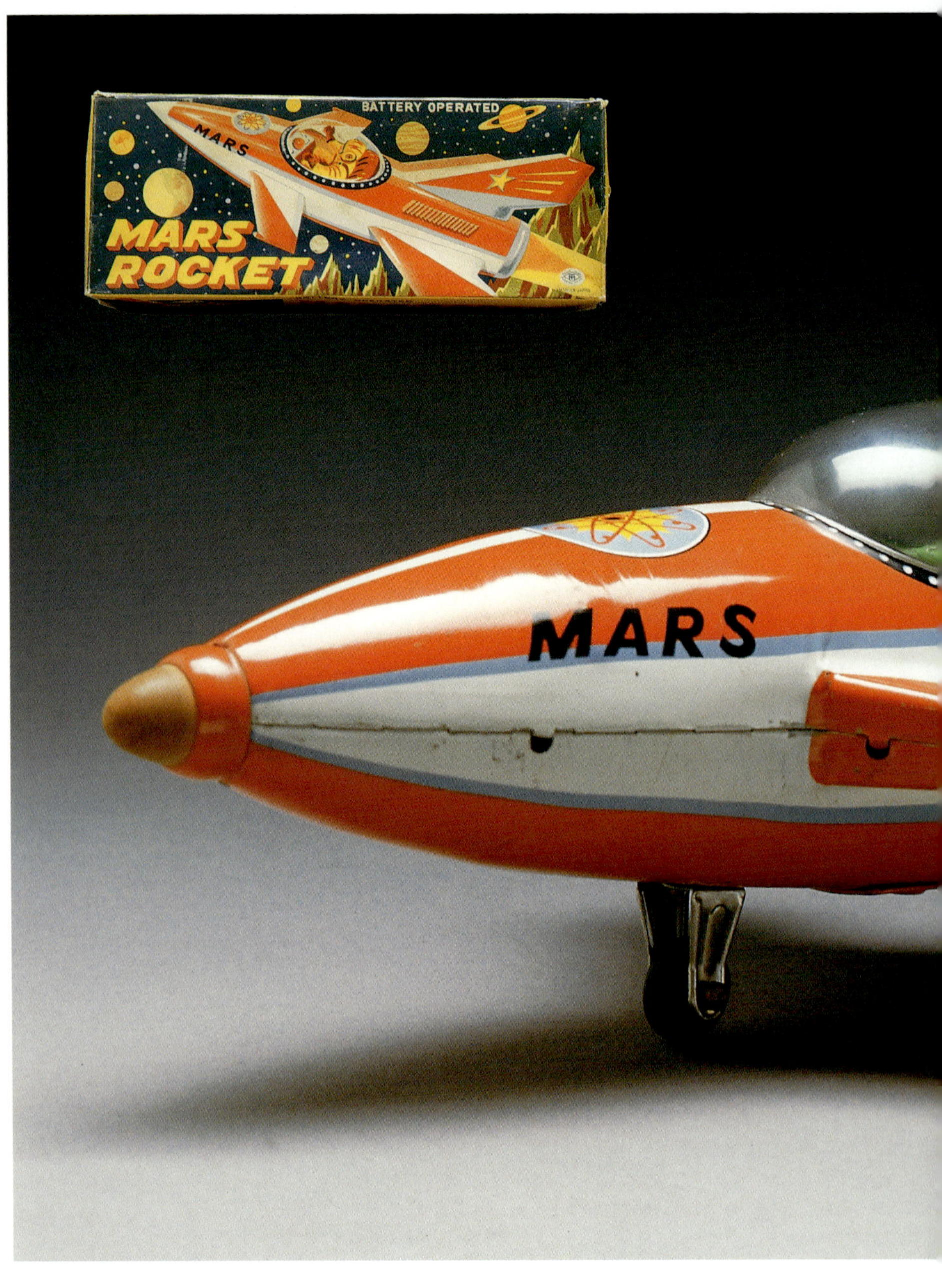

1960s, *Mars Rocket*, Masudaya, 365 x 190 x 140 mm

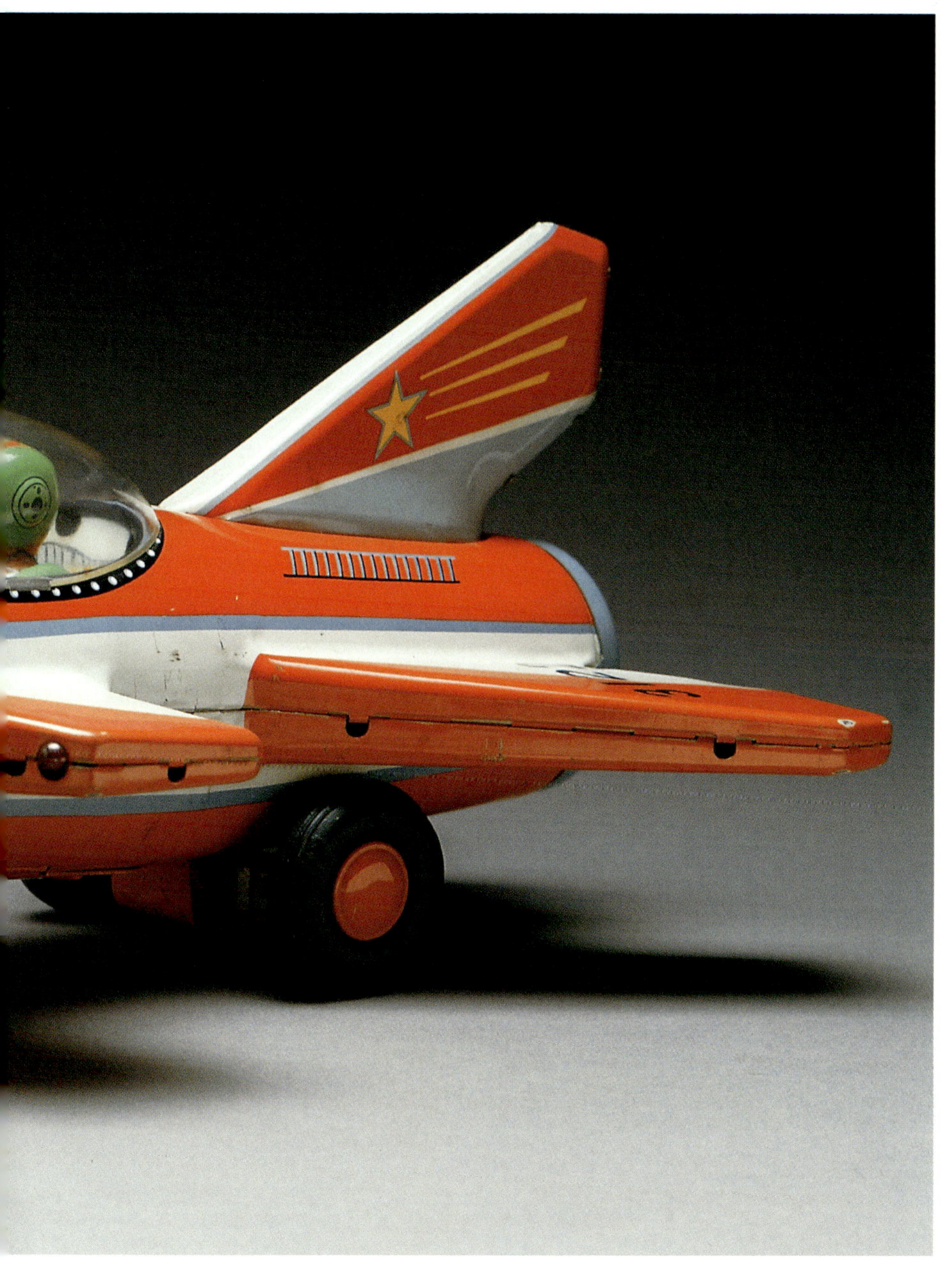

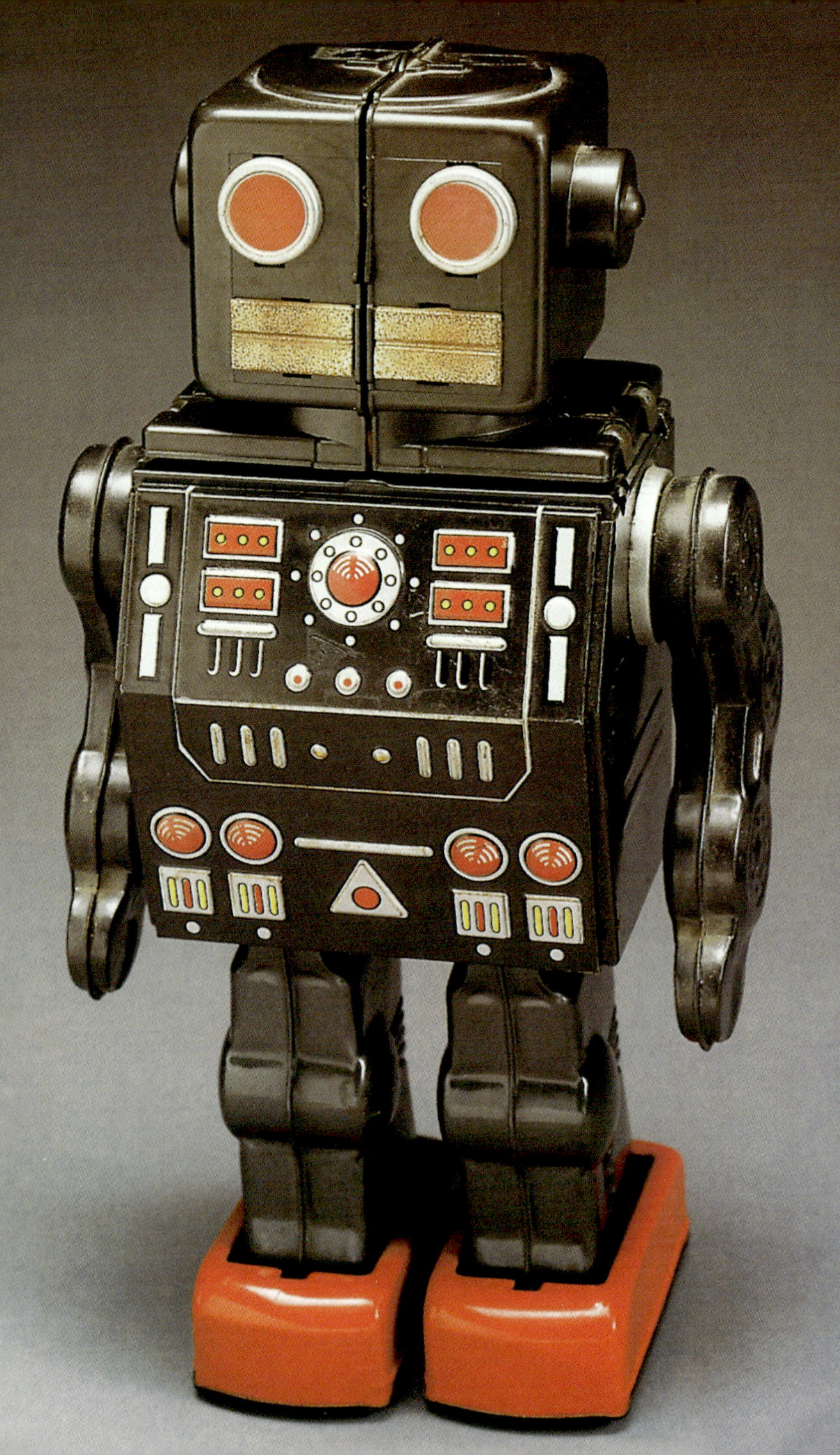

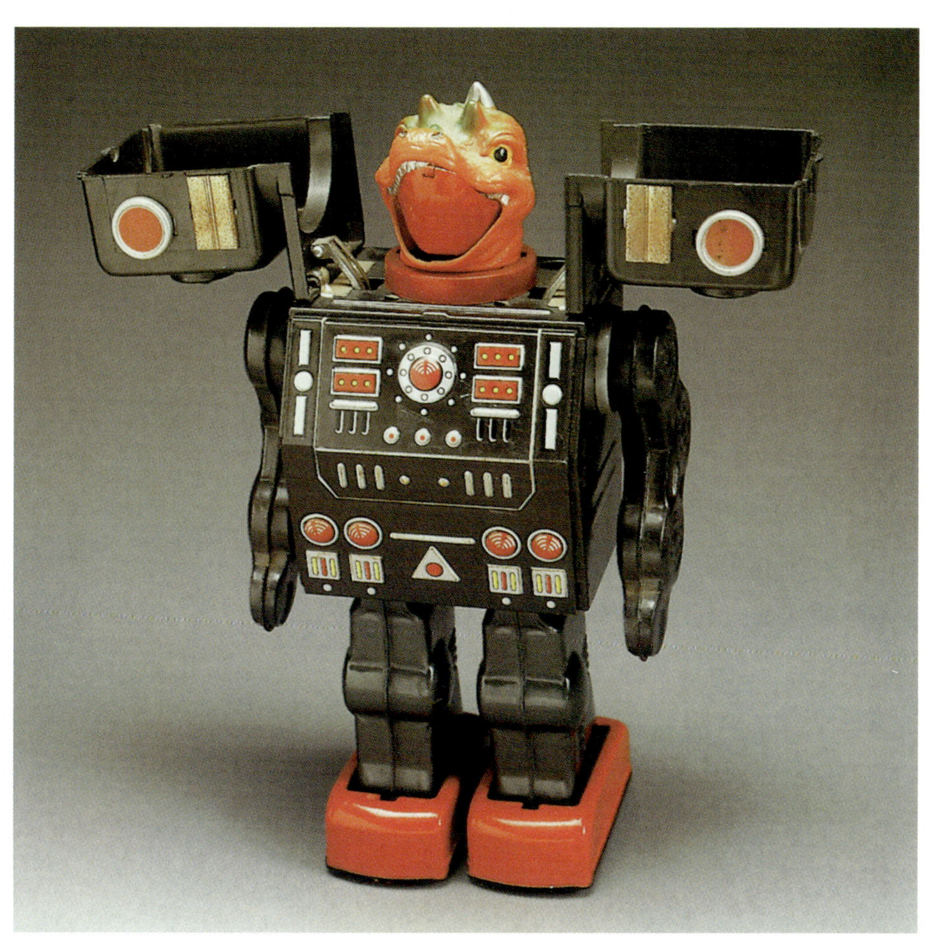

1960s, *Metamorph*, Horikawa, 98 x 140 x 283 mm

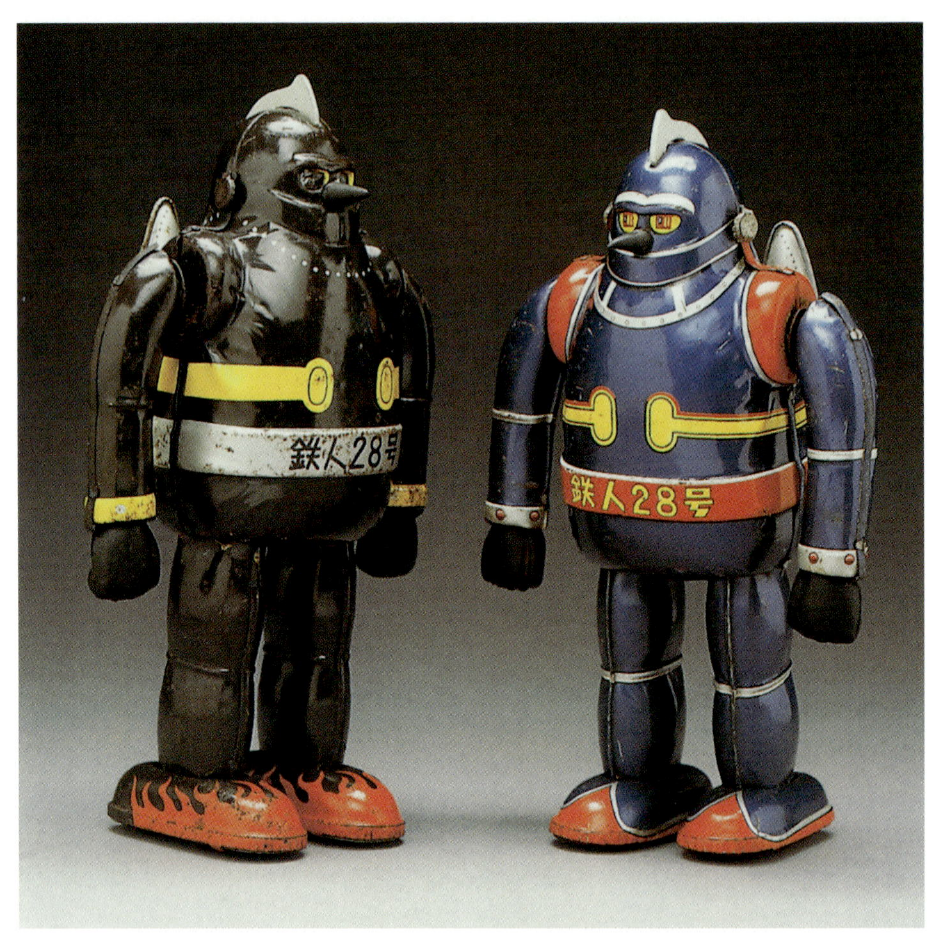

1960s, *Tetsujin 28-GO Robots*, Nomura, 65 x 115 x 195 mm

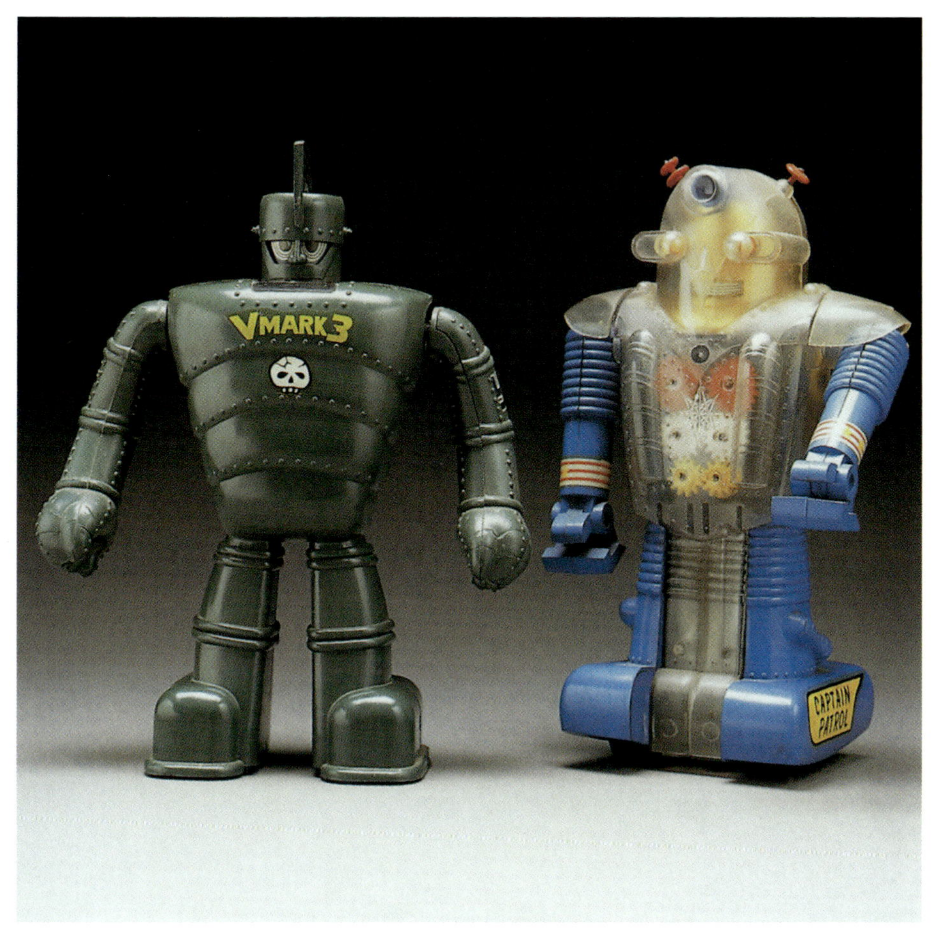

1960s, *V Mark 3*, *Captain Patrol*, unknown, Imai, 65 x 118 x 215, 85 x 85 x 215 mm

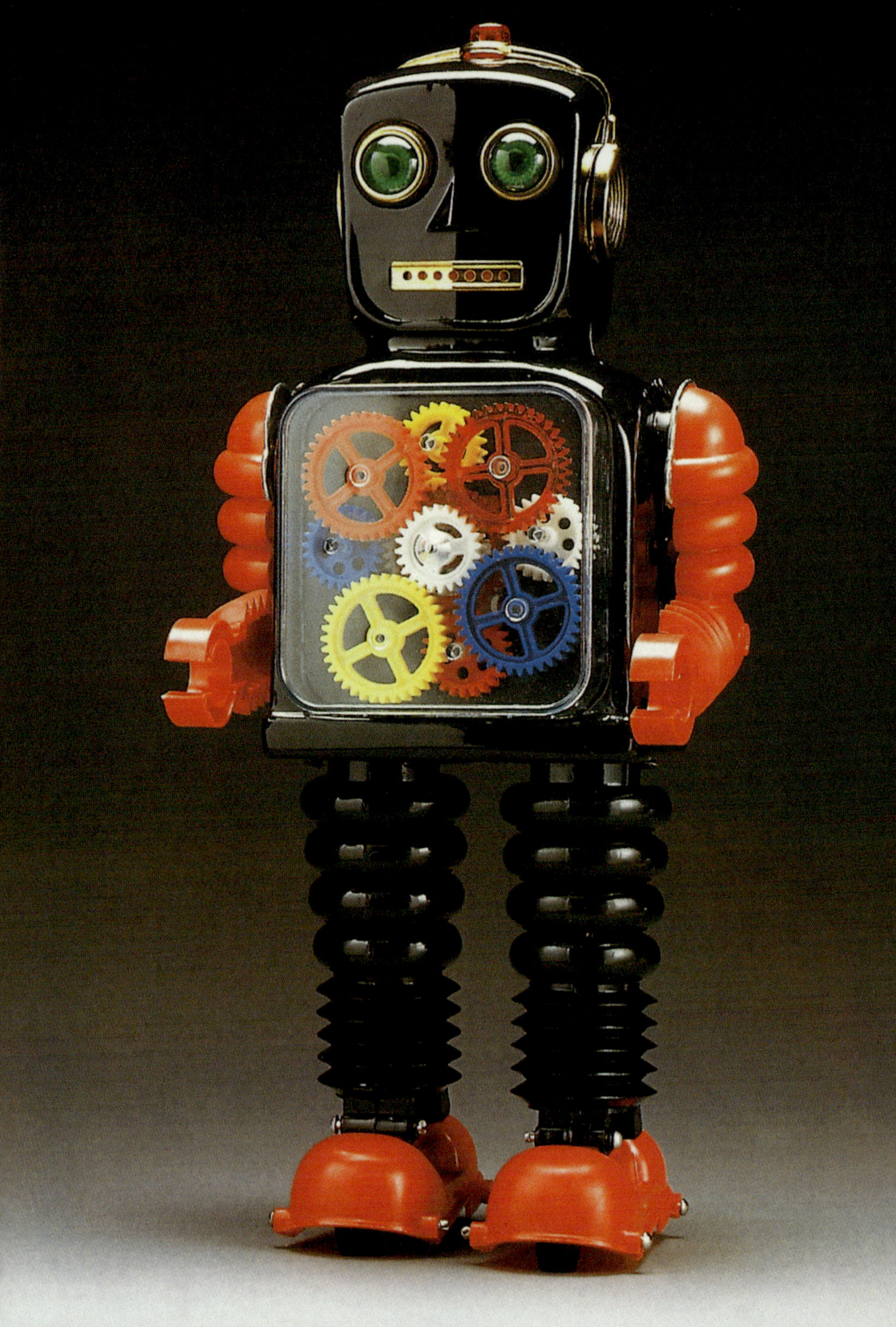

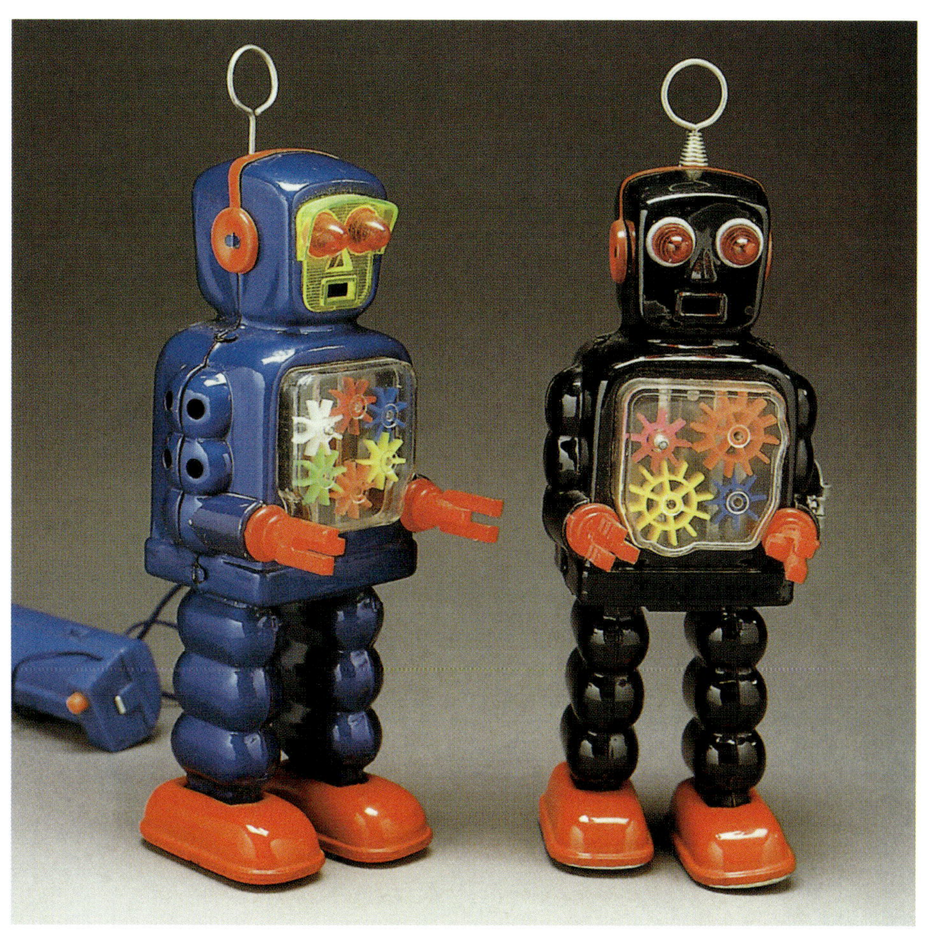

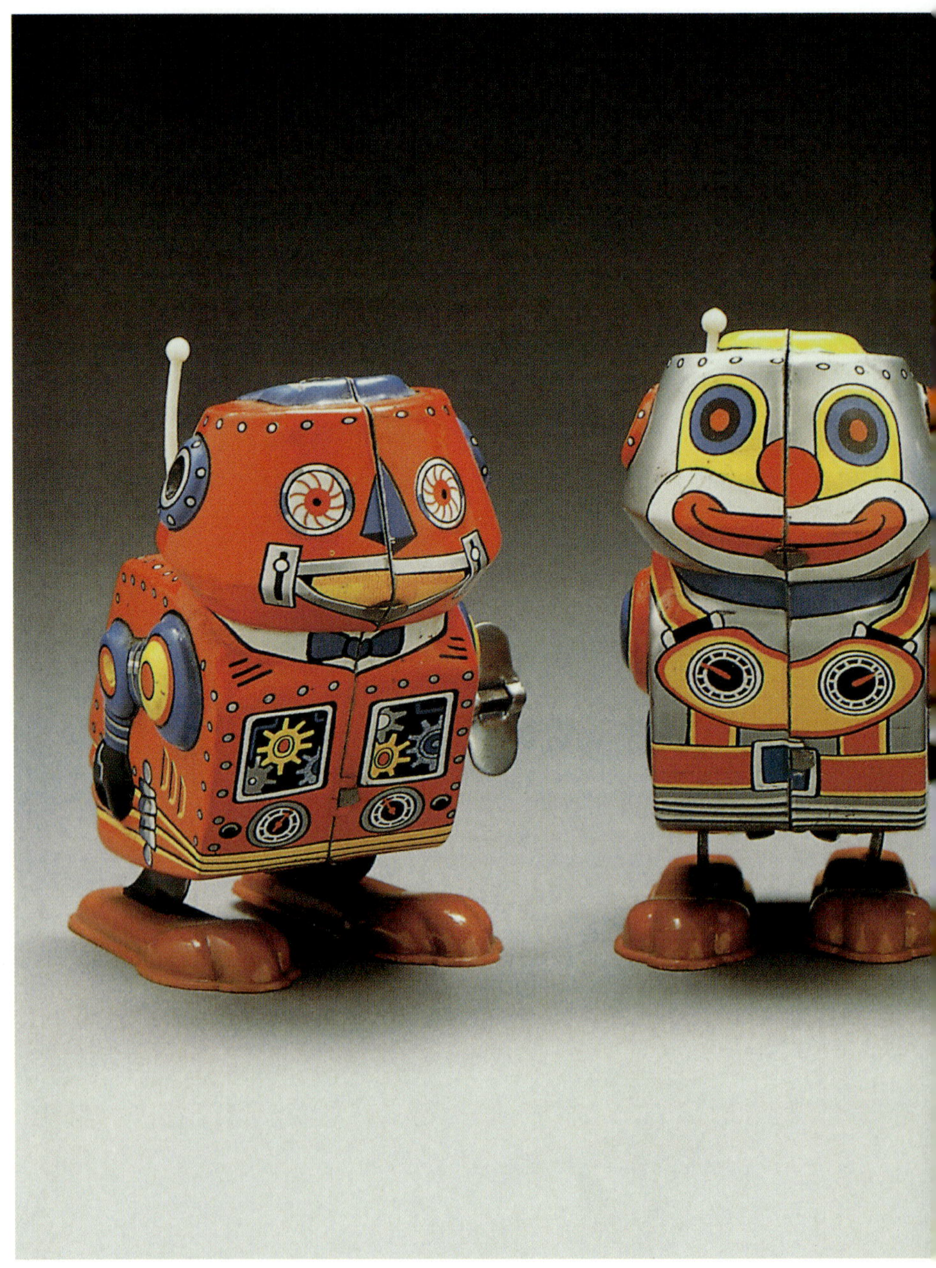

1970s, *Robots*, Yone Toy, 80 x 55 x 85, 80 x 55 x 85, 185 x 55 x 85 mm

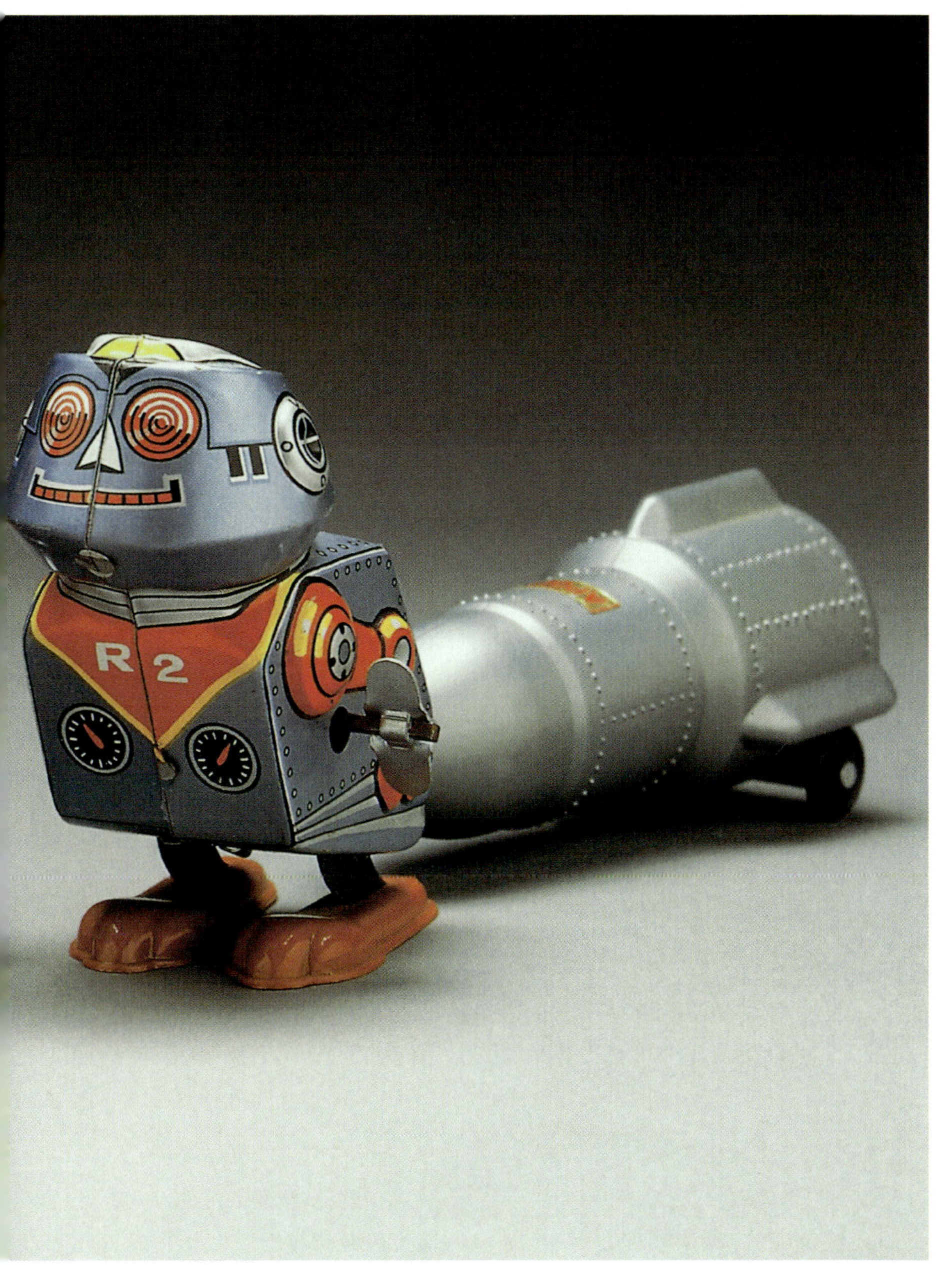

R 2

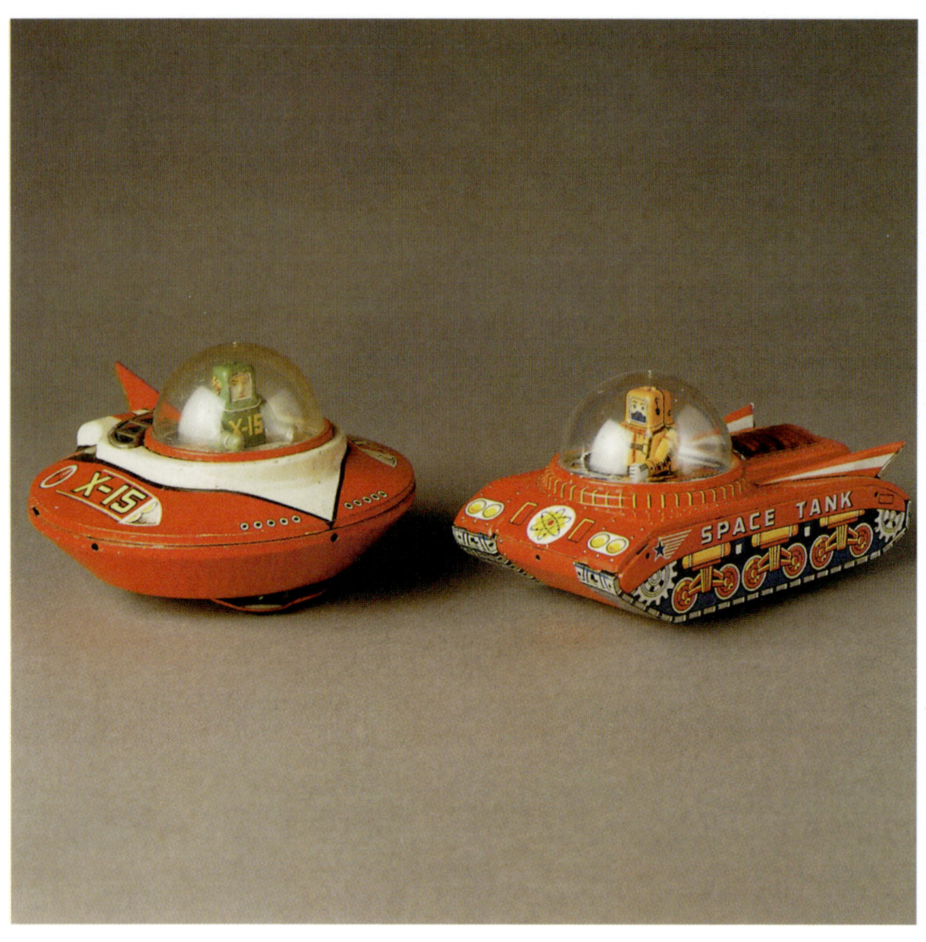

1950s, *Space Patrol*, *Space Tank*, unknown, 140 x 140 x 95, 157 x 90 x 115 mm

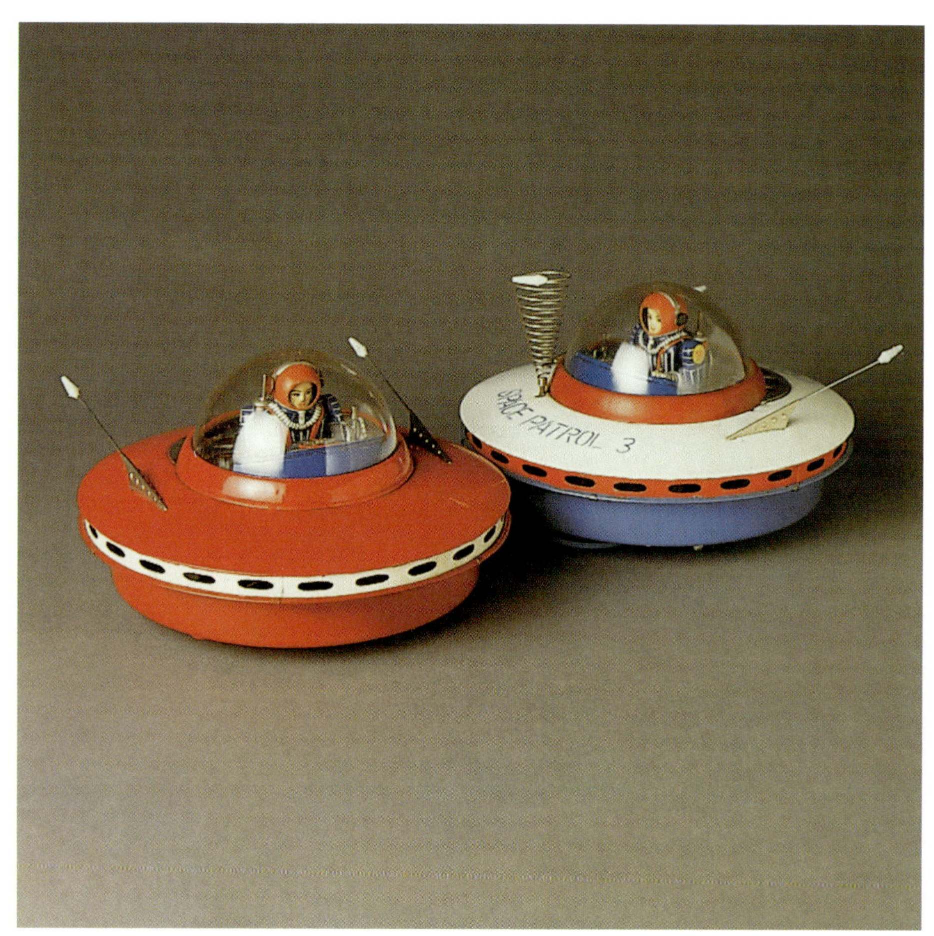

1960s, *Space Patrols*, Yoshiya, 190 x 190 x 120 mm

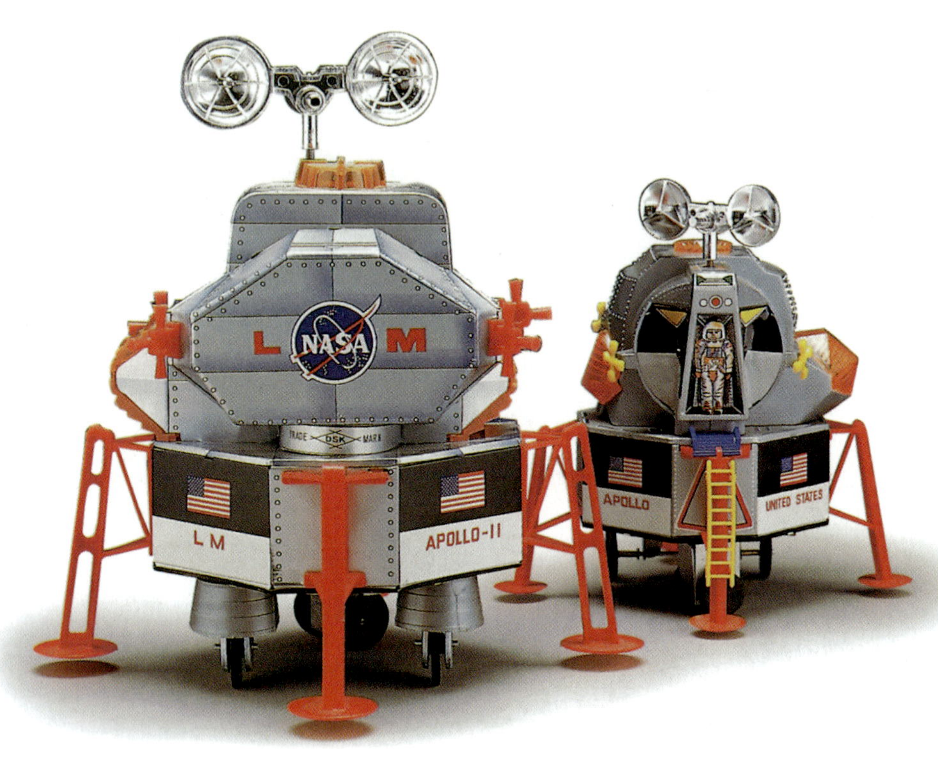

1960s, *Apollo-11 LM*, Daishin, 180 x 180 x 245, 135 x 135 x 183 mm

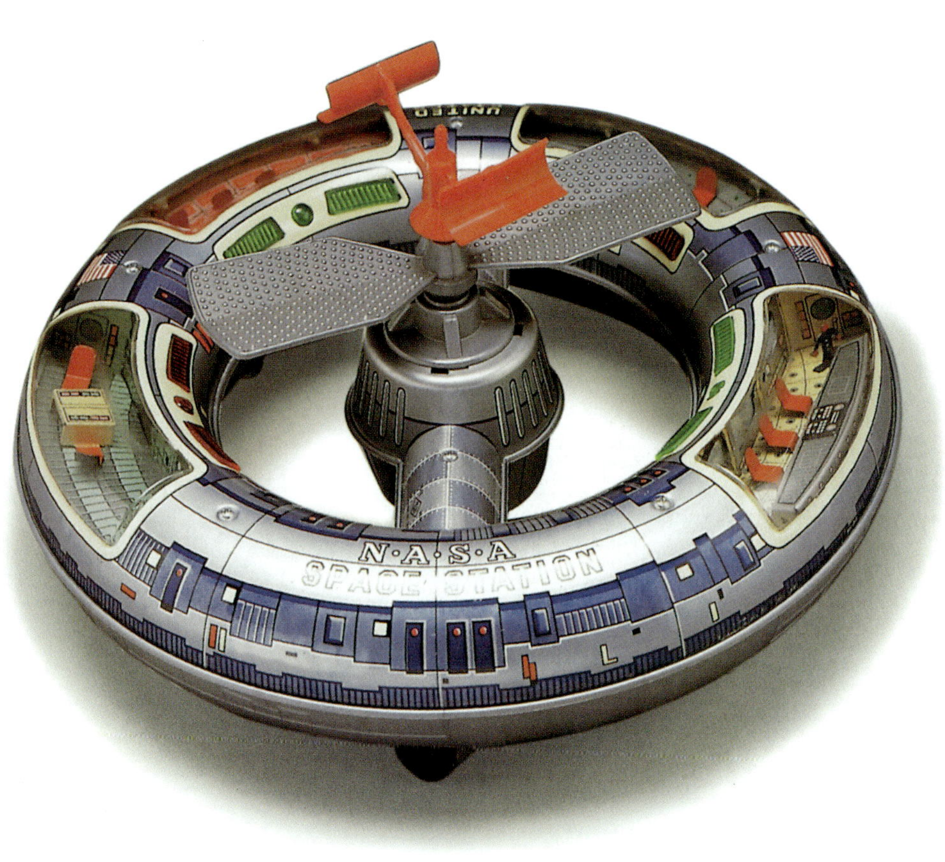

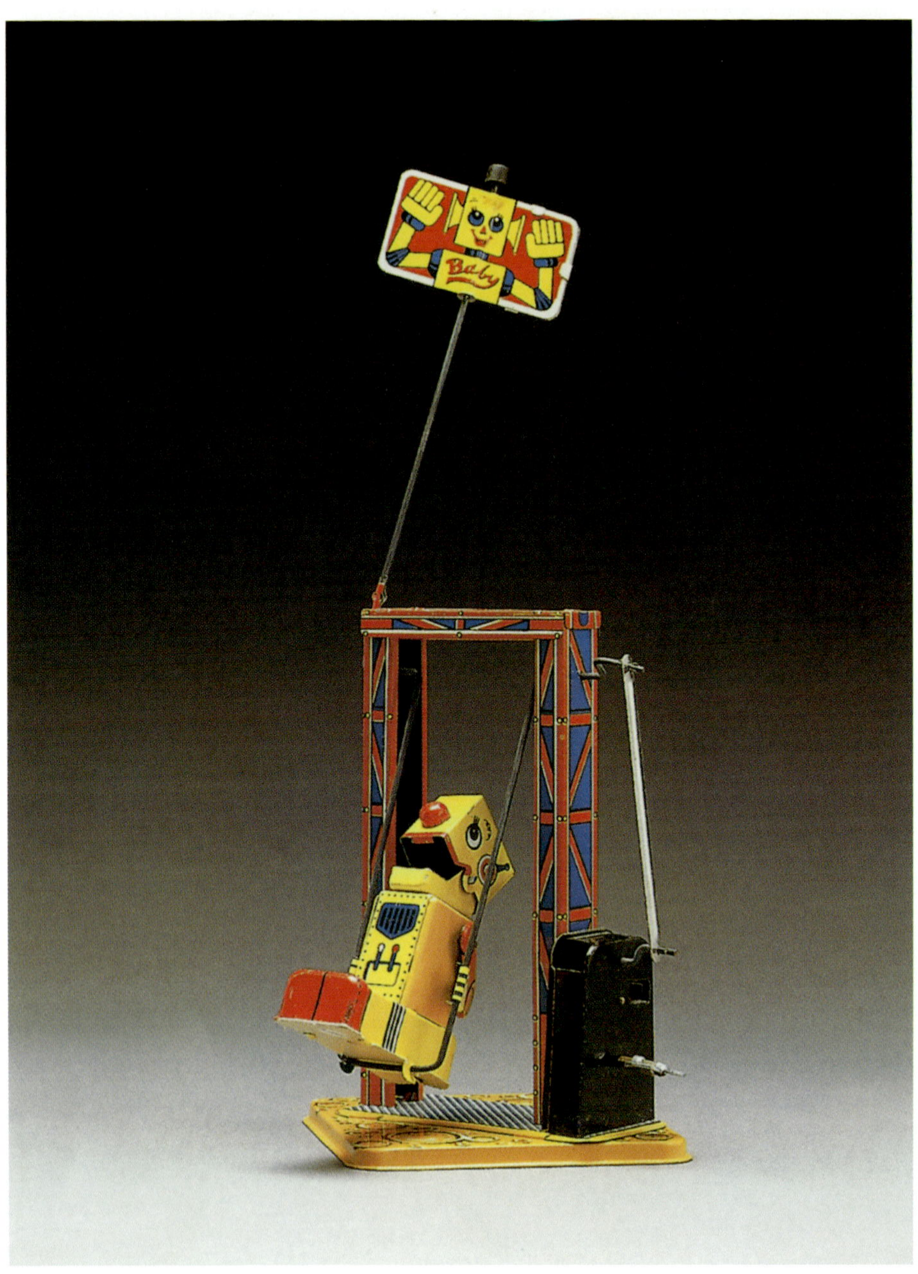

1960s, *Robot on Swing*, Yonezawa, 135 x 135 x 160 mm

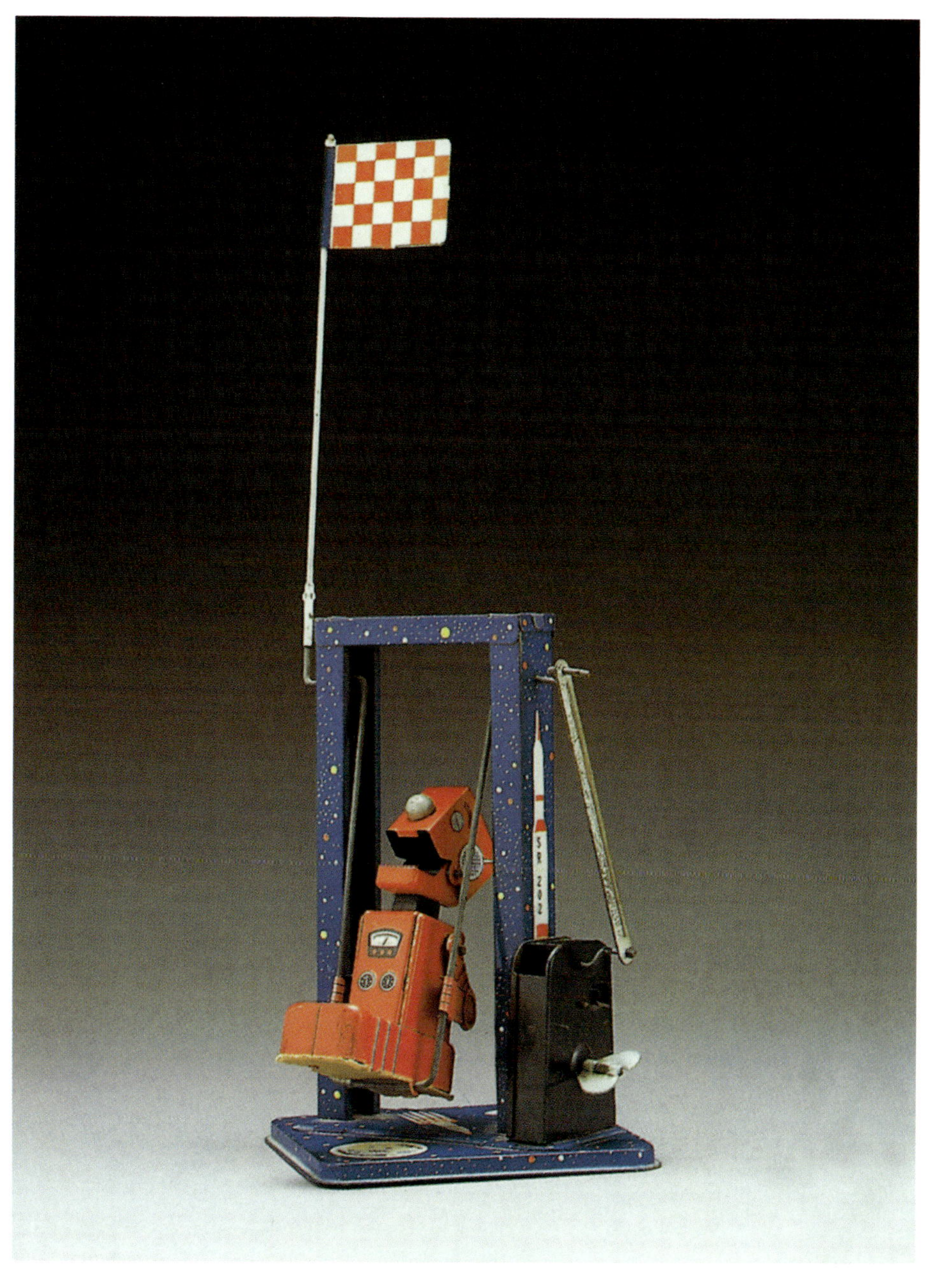

1960s, *Robot on Swing*, Yonezawa, 135 x 135 x 160 mm

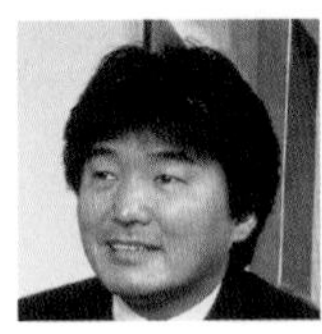

Teruhisa Kitahara

Born in Tokyo. Graduated in economics from Aoyama Gakuin University. He is one of the world's most famous Tin Toy collectors, and Managing Director of Toys Co. He has established 7 Tin Toy museums throughout Japan, including the one near to the Foreign Cemetery in Yokohama. He appears in the popular Japanese TV show, "Kaiun! Nandemo Kanteidan" (We appraise your hidden treasure). His engagements as a popular public figure and connoisseur take him all over Japan.

Geboren in Tokio. Wirtschaftsdiplom an der Universität Aoyama Gakuin. Er gehört weltweit zu den bekanntesten Sammlern von Blechpielzeug und ist geschäftsführender Direktor von Toys Co. Verteilt über ganz Japan gründete er sieben Museen für Blechpielzeug, wie zum Beispiel in der Nähe des Ausländerfriedhofs in Yokohama. Er tritt in der beliebten japanischen Fernsehshow „Kaiun! Nandemo Kanteidan" (Wir schätzen ihr verborgenes Vermögen) auf. Als Fachmann und Kenner mit hohem Bekanntheitsgrad ist er ständig unterwegs in ganz Japan.

Né à Tokyo. Diplômé en sciences économiques de l'université Aoyama Gakuin. PDG de Toys Co, et l'un des plus importants collectionneurs de jouets en fer blanc au monde, Kitahara a créé sept musées du jouet en fer blanc à travers le Japon, dont celui qui jouxte le cimetière étranger de Yokohama. Il est l'animateur de l'émission japonaise populaire « Kaiun ! Nandemo Kanteidan » (Nous estimons vos trésors cachés). Expert jouissant d'une grande renommée, il sillonne sans cesse son pays.

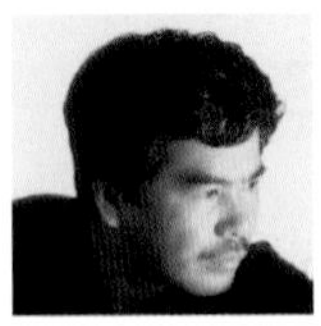

Yukio Shimizu

Born 1944 in Tokyo. Graduated in photography from Art Center College of Design. Active in various areas of art, including commercial films and magazines. Yukio Shimizu strives to capture the essence of his subjects by means of accurate, intelligent images.

Geboren 1944 in Tokio. Fotografiestudium am Art Center College of Design. Auf verschiedenen künstlerischen Gebieten tätig, unter anderem Arbeiten für Werbefilme und Zeitschriften. Yukio Shimizu versucht, in präzisen und intelligenten Bildern die wahre Natur seiner Themen einzufangen.

Né en 1944 à Tokyo. Diplôme de photographie à l'Art Center College of Design. Travaille dans différents domaines artistiques, y compris le cinéma publicitaire et les magazines. Yukio Shimizu s'efforce toujours de saisir la véritable nature de ses sujets par ses photographies précises et intelligentes.

Mr. Kitahara's Toy Museums
Die Spielzeugmuseen von Herrn Kitahara
Les musées des jouets de Mr. Kitahara

Yokohama Buriki no
Omocha Hakubutsukan
Tin Toys Museum of Yokohama
239 Yamatechoh, Naka-ku,
Yokohama-shi,
Kanagawa, 231-0862
Tel 045-621-8710

Kikaijikake no Omochakan
Clockwork-driven Toys Museum
Marine Tower 3F, 15 Yamashitachoh,
Yokohama-shi, Kanagawa, 231-0023
Tel 045-641-1595

Sekai Kyakusenkan
World Passenger Boat Museum
Hikawamaru sennai, Ymashita Koenchi
saki, Yamashitachoh, Naka-ku,
Yokohama-shi, Kanagawa, 231-0023
Tel 045-641-4361

Hakone Omocha Hakubutsukan
Toys Museum of Hakone
740, Hakonechoh Yumoto, Ashigara
Shimo-gun, Kanagawa, 250-0311
Tel 0460-6-4700

Kitahara Kenkyujo Shimizu Omochakan
Kitahara Laboratory,
Toys Museum of Shimizu
S-PULSE Dream Plaza 3F, 13-15
Irifunechoh, Shimizu-shi, Shizuoka,
424-0942
Tel 0120-573-710

Yokohama Character Museum
The Character Museum of Yokohama
CrossGate 4F, 1-1-67, Sakuragichoh,
Naka-ku, Yokohama-shi, Kanagawa,
231-0062
Tel 045-683-3155

Kitahara Kanteidoh Hanjoh
Nandemo Hompo
Connoisseurship Hall of Kitahara
inside Namco NanjaTown, Sunshine
World Import Mart 2F, 3-1-3 Higashi
Ikebukuro, Toshima-ku, Tokyo,170-0013
Tel 03-5992-2545

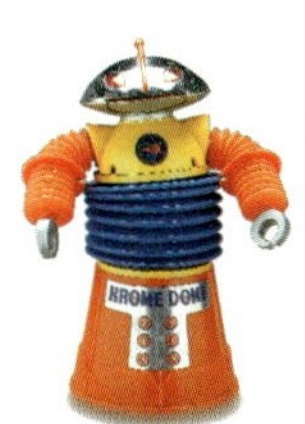

http://www.toysclub.co.jp